BUSINESS/SCIENCE/TECHNOLOGY DIVISION
CHICAGO PUBLIC LIBRARY
400 SOUTH STATE STREET
CHICAGO, IL 60605

THE COMING ENERGY REVOLUTION

JEANE MANNING

Avery Publishing Group

Garden City Park, New York

Cover design: William Gonzalez and Rudy Shur
Cover photo: Peter Gregoire/Index Stock Photography Inc.
In-House Editor: Lisa James
Typesetter: Bonnie Freid
Printer: Paragon Press, Honesdale, PA

Library of Congress Cataloging-in-Publication Data

Manning, Jeane.
    The coming energy revolution : the search for free energy / by
Jeane Manning.
      p. cm.
    Includes bibiliographical references and index.
    ISBN 0-89529-713-2
    1. Energy development.   2. Power resources—Environmental aspects.
I. Title.
TJ163.2.M365 1996
621.042—dc20   96-1135
                                           CIP

Copyright © 1996 by Jeane Manning

All rights reserved. No part of this publication may be reproduced,
stored in a retrieval system, or transmitted, in any form or by any
means, electronic, mechanical, photocopying, recording or otherwise,
without the prior written permission of the copyright owner.

Printed in the United States of America

10 9 8 7 6 5 4 3 2 1

RO1166 50131

BUSINESS/SCIENCE/TECHNOLOGY DIVISION
CHICAGO PUBLIC LIBRARY
400 SOUTH STATE STREET
CHICAGO, IL 60605

Contents

Part IV The Energy Revolution—Potential Amid the Problems

*To the pioneering inventors in the new-energy scene,
and especially to all those whose stories did not
end up in this overview—your struggles and
triumphs nevertheless touch and
inspire me*

Acknowledgments

I would like to thank all those who have contributed to this book. It is impossible to name each person and fit the list onto a page, but in my heart and mind the thought of each individual researcher, family member, supportive friend, helpful scientist, and inventor is surrounded by my gratitude.

A more formal acknowledgment is owed to the Canada Council for the Arts, Explorations program, for its encouragement in the form of finanacial assistance while I wrote my unpublished autobiographical manuscript, *Beyond Tesla*. This book grew out of that work.

Last but not least, I want to thank Lisa James, my editor at Avery Publishing Group Inc., for good-humored patience and for constantly reminding me that this book is written for the public—people who may or may not care about technical matters but who do care what a new energy technology would mean to their everyday lives.

Foreword

This may be one of the most important books you will read. It describes the rapid progress in making available a source of energy that many of us in the scientific community believe will radically change the face of the earth—zero-point energy from quantum fluctuations in the vacuum of space. Many people call this "space energy" or "free energy."

Because the free energy that surrounds us is such a vast source, so potentially clean and decentralized, some of us believe a revolution is brewing. Afterward, existing energy sources may be seen as dinosaur nightmares that littered our landscape and polluted our air. This revolution could open a new paradigm of science and technology that would make the Copernican and Industrial Revolutions appear tame. Redirecting $2 trillion per year from obsolete power systems to sustainable and affordable systems is unprecedented.

Electrical power systems based on compact solid-state devices will probably replace the fuse boxes and circuit breakers of individual homes and buildings. We finally will be able to get off the power grid. The new energy boxes will also be portable, eliminating the need for storage units such as batteries. They will replace the internal combustion engines in our automobiles and other transportation and industrial systems. And they can be used in the field to dramatically increase agricultural efficiency (for example, pumping for irrigation) and therefore can help eliminate hunger in Third World countries.

But we will need to develop policies in which these energy devices are scaled appropriately to their end use, and not overused or applied to weapons. In the longer term, I believe we will discover that space energy can also be used in a practical way for anti-gravity propulsion systems.

So why don't we get on with it? It seems that since the time of Nikola Tesla a century ago, we have been suppressing "free energy." We have all developed some deep-seated fears that are blocking the way, ones that haven't yet reached the public consciousness. Yet with all the suppression, the energy genie is finally out of the bottle, as Jeane Manning clearly shows in this book.

First, I'd like to share some background to these dramatic statements. About twenty years ago, when OPEC raised its oil prices and an energy crisis erupted, I began to look at how we in our culture were abusing energy.

Do you remember the gas lines of the mid-1970s, the grim statistics of increasing oil scarcities, air pollution, oil spills, oil wars, prophecies of nuclear meltdowns, radioactive waste contamination, nuclear weapons materials proliferation from foreign domestic power programs (like North Korea's now), and other predictions of doom and gloom? As we head toward the turning of the millennium, whatever happened to the energy crisis? Whatever happened to the Club of Rome report on the limits to growth? Many of us can recall that as the 1970s further unfolded, the OPEC cartel began to lose its grip, oil prices dropped, and Ronald Reagan was elected U.S. president.

As if by magic decree, the mass media and public consciousness decided there was no energy crisis after all. The oil glut resumed and any perception that there was a need to develop alternatives seemed to drop out of public awareness.

During 1975, I was special consultant on energy to former Congressman Morris Udall's subcommittee on Energy and the Environment of the U.S. House Interior Committee. I wrote speeches and orchestrated hearings for eight months for Udall while he was running for president. I helped him develop an energy policy not unlike what President Jimmy Carter began to implement during the late 1970s, and continued by the Clinton administration.

The policy acknowledges the grim statistics of a highly polluting and expensive energy future in the coming decades if we do not at least slowly move away from fossil fuels and nuclear power as the mainstays of our electrical power generation. We are also bringing in strict emission controls. Yet underlying these positive steps has been an enormous blind spot and a resistance to innovative solutions to the ongoing worldwide crisis.

Even some of the most innovative technical and policy organizations in the energy/environment field, such as the Rocky Mountain Institute and the Union of Concerned Scientists, seem to be resigned to slow and modest changes—more use of traditional renewable sources (solar, wind, hydrogen, biomass, and others) and increased automobile efficiency are incremental reforms. In the long run, neither the limited supply of fossil fuels nor the declining quality of the environment can be swept under the rug for much longer. The problems of the 'seventies have become the even greater challenges of the 'nineties.

Two decades ago, the fleeting public perception that we did have an energy challenge helped to spawn a Department of Energy (DoE). It was expected to support research and development in alternative energy sources. But much of the DoE was simply bureaucratic old wine in new bottles, combining existing special interests in both fossil fuel and nuclear power.

Until this day, it is apparent to me that little true progress has been made to stop our abuse of energy and the environment. Rather than moving Manhattan Project- or Apollo Project-style into the future, we instead reinvent the wheel of vested interests in fossil fuels and internal combustion engines and nuclear energy, and we continue to supply electricity from large central power stations though an ugly grid system that may be a major health hazard (electromagnetic pollution from power lines). So why didn't we do anything about this?

A switch to clean "free energy" could almost totally alleviate air pollution, global warming from carbon dioxide emissions, waste heat, Saddam Hussein's ecocidal fires, black skies, oil spills, acid rain, nitrogen dioxide, sulfur dioxide, hydrocarbon and ozone emissions, unsightly oil production and refining facilities, supertankers, gas stations, power stations, transmission lines, and the rest of it.

Use of "free energy" could also end our thirst for oil and natural gas. This thirst is draining precious resources from the Earth at alarming rates. This lifeblood, formed over tens to hundreds of millions of years within Earth's crust, has been greedily extracted as if there were no tomorrow. Oil production and consumption has more than tripled since the onset of the energy crisis. Almost half of the world's available oil and more than half of the natural gas have already been skimmed off the top of our best deposits and burned, mostly within one human generation!

At present rates of consumption, proven U.S. oil reserves will last just ten years, and world oil reserves will last forty years. Even if these reserves were to prove to be twice as abundant as the estimates, we will run out of oil by the middle of the twenty-first century, with inevitable sharp price rises.

These facts have been ignored by our energy policy-makers. We are indeed borrowing the Earth from our children rather than inheriting it from our parents. An ecological consensus is emerging—*we must stop this and build a sustainable future.*

The economic impact of converting to space energy would be enormous. Revenues from the use of electrical power worldwide are now $800 billion per year, a doubling over the twenty years that have passed since awareness of the energy crisis was articulated and then withdrawn. This staggering cost even exceeds by twofold the size of the automobile industry, and is comparable to the amount that taxpayers annually pay to a debt-ridden United States government.

The worldwide energy infrastructure that depends primarily on burning oil, coal, and natural gas, and on the use of radioactive elements, consumes about $2 *trillion* each year, a figure so high it is hard to imagine the enormity of its grip on all of us. In the time it takes for you to read this sentence, the world is burning up more than one million dollars of fossil and nuclear fuels for use in electrical appliances, heating, cooling, and transportation systems.

During the early 1980s, while I was studying advanced space power concepts at Science Applications International Corporation, it became very clear to me that any radical new idea in the energy field was in for tough sledding. It would face vested interests within the U.S. government and established industry.

Most of the billions of dollars of Department of Energy research and development funds are still spent each year to expand the use of fossil fuel and nuclear energy. In my years as a science policy analyst, I learned that government R&D projects form the thin edge of a wedge of great political and economic clout; today's blueprints leverage into tomorrow's multibillion dollar realities. Once a project's investment goes over a billion dollars, the project becomes a new special interest, with contractors in Congressional districts and so forth. This guideline appears to hold regardless of the merit of the project.

The largest single advanced R&D project in the DoE is the more than a billion dollars spent on the (still infeasible) "hot" fusion concept. Hot fusion would involve both building large power plants and more pollution from excess heat, radiation, and power lines. Another significant portion of DoE funds is spent for high-energy physics and weapons research not directly related to energy production. Much smaller amounts go to developing solar and other alternative sources, and to energy conservation. Nothing—not one penny—of American public funds (outside the black budgets that we don't know about) is invested in looking at the source of energy that I believe will change the way we do things—the free energy which surrounds us.

An entrenched interest has become so powerful that we seem to be blind to any new concepts, especially those as radical as "free energy" and cold fusion. We seem to be more interested in the controversy about whether these developments are real, rather than in seizing a golden opportunity.

In short, we seem to have sunk into a false sense of security, continuing with an abusive energy infrastructure that is destroying the Earth and ourselves. We have created for ourselves an "electric jail," being increasingly boxed in by a grid of unsightly, unhealthy power lines and gas stations, and the endless droning of internal combustion engines and other energy-related facilities that litter the landscape and the skies and the oceans.

Like the frog who is gradually cooked in a pond where the temperature is gradually raised, we have gradually acclimated ourselves to our electric jail. Gridlocked within, we forget how enriching the more sustainable environments of even twenty years ago felt to us. I live in the remote woods of the Oregon Cascades and can attest to the benefits of a peaceful, clean environment.

Most of us have not faced the fact that we have abused our energy resources so badly and in such a short flash of history. We must eliminate this Earth-battering if we are to survive.

Much theoretical and experimental information already supports the credibility of space energy. I have been surprised to see a breadth and depth of knowledge, dedication and professionalism, and substantial achievements among leading theoreticians, experimenters, and inventors in the "free energy" field. These are the explorers of a new reality. They are cut off from the mainstream because the mainstream debunks this reality, with a denial based on the most superficial reasoning.

Rather than the public's stereotypical image of the eccentric out-of-touch garage inventor who is probably wrong, many of our "free energy" inventors and researchers are Ph.D.s working in mainstream settings, such as Shiuji Inomata at the Electrotechnical Laboratories in Tsukuba (Japan's "Space City"). He has been a full-time government employee for the past thirty-five years.

In India, Paramahamsa Tewari has a prestigious government position as Chief Project Engineer of that nation's largest nuclear power plant under construction. Both governments have permitted these two men to build their "free energy" devices (based on Bruce DePalma's concept), something that has been unthinkable in a DoE lab in the United States.

Dr. Inomata recently lectured government and industrial leaders (more than 600 professionals showed up at his last seminar), and Toshiba Corporation invested $2 million to develop superconducting magnets for his new unipolar generator. Being almost totally dependent on foreign oil for its energy and transportation needs, Japan has little to lose and a lot to gain from commercializing the free energy of surrounding space. This could be another opportunity missed by other nations, with even more potential Japanese market dominance of products resulting from their farsighted approach. When will we ever learn?

Few other qualified scientists bother to take the learning and relearning time to learn about "free energy"; most of the vocal naysayers have not addressed the puzzle. They are limited by peer pressure and funding pressures, and by a strong bias against probing the unknown outside their own specialties. I know; I was there!

A common error made by the debunkers is the assumption that if these machines were real, they would have heard about it. The history of science is full of examples of leading scientists ridiculing—sometimes emotionally—new ideas because of this assumption, and later being shown to be wrong. Of course this is about as far from science and rationality as you can get, and it suggests that the suppression syndrome starts with scientists themselves.

Government officials and the media turn to the scientists for their information, and so also ignore the obvious. For example, a *Washington Post* reporter interviewed some of us at a New Energy symposium on the topic of space energy (also known as free energy). Instead of the in-depth briefing we gave the reporter, what appeared in the newspaper was a light-hearted "safe" historical piece about Nikola Tesla.

By default, it seems, these establishment mouthpieces define what is meant by credibility—which may actually have little to do with the truth. One phone call from a mainstream journalist to a mainstream luminary such as Carl Sagan, for example, could quash a story. Unfortunately, our most revered news sources do not have the final word on the truth, and this causes the slow progress of science.

So why haven't we adopted "free energy" if we've had it for so long? Why have we gone to war over oil—in the Gulf War, Somalia, and even in Vietnam? Why have we misdirected untold trillions of dollars and sacrificed human lives and our surroundings, and why do we continue to do so in the face of evidence that we can stop doing these self-defeating things?

How could so many decision-makers have kept "free energy" so completely from us, so that there is still not a single machine on the market? After all, the technology for making it available is probably not that far beyond our reach. It is probably much less challenging of a project technically and financially than the Manhattan Project scientists faced in developing the atomic bomb, or than the Apollo program scientists and engineers faced in sending men to the moon, or than the Ph.D.s face working at the Princeton tokomak hot fusion project, which is still far away from the elusive "breakeven" point. Inventors have apparently been demonstrating "free energy" results for mere thousands of dollars—not the billions and trillions spent on perpetuating more traditional approaches. Why has it taken so long for money to flow in logical directions?

In other words, how could the suppression of "free energy" technology have been so complete, so airtight for so long? If our governments and scientists are ignoring the obvious, why haven't market forces gotten wind of this and briskly moved ahead? It seems that everybody is waiting for the other shoe to drop.

I have come to a conclusion I had previously thought to be unlikely: that the Suppression Syndrome pervades every aspect of any revolutionary new development. Usually the more radical the concept, the stronger will be the forces of suppression.

For example, most inventors are underfunded or have been "bought out" in exchange for keeping their trade secrets under wraps. This closes them off from sharing knowledge within the interdisciplinary teams that I feel will become necessary to develop this new industry. In my opinion, we will need a moderately funded effort of perhaps tens of millions of dollars to make the necessary breakthroughs.

I disagree with those who see this potentially paradigm-shattering development as a purely competitive private-sector issue—a horse race motivated by the chance that a particular system might be the winner which could yield millions or billions of dollars to lucky investors. In such a competitive situation, other new energy systems fail to be developed because of bad timing or underfunding or other suppressions.

In such a win-lose system, we are spinning our wheels. In the Western world, the entire complex of denials from scientists and secrecy from industry is gridlocking us. It's a crazy system!

In summary, most inventors and researchers of "free energy" systems are underfunded, so progress is slow. The prospect of becoming a millionaire by being among the first to develop a commercial model encourages secrecy and suppression. Instead of this all-or-nothing approach, I propose that we develop win-win funding strategies that would virtually eliminate the cancer of suppression. (Because of our fear of the unknown, we are suppressing what we need the most.)

There appear to be three main challenges for "free energy" proponents:

1. *Suppression* of all kinds has been efficiently blocking availability of the new energy technologies.
2. The potential of "free energy" for replacing existing infrastructures will cause *displacements* in jobs, income, and power, to a degree that is unprecedented in our economy.
3. The *abuse* of "free energy" technology could lead toward its overuse or use as a powerful weapon. However, devices can be designed to be safe. I feel we cannot let the potential for abuse be a reason to stop or to suppress the technology.

Harnessing the clean "free energy" is too important for the planet and for ourselves, and is inevitable. But we must develop standards for appropriate use, to meet the strictest guidelines for sustaining our global environment. We need to be responsible creators—learning lessons from our abuses of nuclear energy, for example.

I do feel that once "free energy" devices pass the usual tests, such as cleanliness, cheapness, and convenience, the technology will quickly flood the worldwide marketplace. We all know of the profound effects that earlier inventions have had on our lives—inventions such as electricity, telephones, automobiles, airplanes, television sets, transistors, and computers, to name a few.

Developments on a near horizon will have an even deeper effect that transcends dollar values. Indeed, Future Shock is here, and most of us appear to be uneducated about the inevitable displacements that can be created by making a multitrillion-dollar industry become obsolete.

Perhaps the greatest suppression of all is our (mostly unconscious) fear of the unknowns that await us on the other side of change. Therefore, we want to deny the change for as long as we can, until the perspective is so obvious we can no longer ignore it.

In our science and technology we are at that watershed time, the time of paradigm shifts, when we decide as a culture to move from one set of truths or realities to another. The old Newtonian view is beginning to go the way of the Flat Earth Society, and yet the prevailing wisdom in our conscious minds is still a Newtonian one.

Along with change comes the emotions of grieving for an old worldview. Well-established research on the stages of the grieving process suggests that soon most of us will move from our current denials toward the stages of anger, bargaining, depression, and finally acceptance of the new.

I am quite certain that as these new energy revelations begin to shake the orthodox world and its delicate economic structures, many of us will experience a great deal of anger or fear—anger about the suppression issue and fear of transition to the new paradigm. A bumper sticker reads, "The truth will set you free but first it will piss you off." Personally, I feel I have moved on to a stage between depression and acceptance.

The challenge is not to decide whether or not "free energy" is real. It is. Instead the challenge is to our collective will, to break free of our ignorance, the electric jail, the ecocide, the gridlock, the Newtonian rigidity, the greed, and the vested interests.

Now I think you can see why I believe this book is so important. "The energy revolution," Jeane Manning said to me in a candid moment, "could affect people's lives—their practical everyday choices—profoundly, because decentralized power means freedom. It means empowerment to clean up our environment instead of feeling helpless. The megaproject-builders now have no leg to stand on when they claim their projects are necessary."

"People have to take back their power at the individual and local level," she continued. "It won't be handed back in an envelope from the government. The tax structure as well as electric utilities will have to be changed, because of all the energy-related tentacles that stretch out from capital cities and from financial centers like Wall Street and reach into citizens' wallets.

"Academia is not always working for the people or their planet. Employees and contractors (mostly of the Department of Defense) scramble for grants and contracts and learn to think in the accepted way in order to stay on the list."

Based on my experience over the past thirty years, I could not agree more. Jeane Manning is a highly qualified journalist who has researched the new-energy scene since 1982. She brings an international perspective to the topic, being in ongoing contact with many inventors, theorists, and other networkers in about a dozen countries. She has attended more than twenty energy-related conferences in Switzerland, West Germany, Canada, and the United States.

She has a bachelor's degree in sociology, and has served as a social worker, reporter, newspaper editor, columnist, and magazine staff writer. Her motivation for leaving the career ladder to pursue such an off-limits topic is her concern for our environment. She has been attuned to nature all her life, having been born in Alaska near the then-pristine Prince William Sound and having grown up in the country near Coeur d'Alene, Idaho. In Colorado, British Columbia, or wherever she lives, Jeane gravitates toward natural settings.

Jeane tells me that she vowed to herself, in 1982 when she first saw an unorthodox magnetic motor, "If this is for real, then I want to tell the public about it when the time is right. We won't need to dam any more wild rivers or poison the air."

The time *is* right to share this with the world. I only hope, for all our sakes, she will get her environmental wish soon. She is the first experienced journalist to cover this important and neglected topic in a trade book, and it fills a unique need.

Another aspect is the fact that Jeane is a woman and a mother. In an industry dominated by men, the feminine perspective is sorely needed as a voice for positive change. If we are to bridge the paradigm gap to make our dreams come true, the job cannot be done merely by those technocrats, scientists, engineers, and traditional media that got us into this fix in the first place. "No problem can be solved from the same consciousness that created it," were Albert Einstein's words.

I hope that you will enjoy reading *The Coming Energy Revolution* as much as I did. Jeane Manning presents an objective view of a feasible technology that is waiting in the wings for its expression—a technology that I believe will lead to a new consciousness on our planet.

Brian O'Leary, Ph.D.
Physicist and former astronaut

Preface

There is growing evidence for a new type of energy which is neither nuclear nor chemical. This has been called zero-point energy.

—Edmund Storms,
Physicist

New ideas are resisted. . . . But we must rapidly explore these new technologies, because what is at stake is life.

—Adam Trombly,
Astrophysicist

There is a fast-growing international effort to completely change the sources of energy on which our world is based. Some of its proponents call it "free energy." Some call it "space energy" or "zero-point energy." By any name, it has the potential to affect the life of every human being on earth.

Long confused with the old, discredited idea of perpetual motion, space energy—the term we will use in this book—is real, as are the other energy technologies we will explore. The existence of these new energy technologies has provoked strong opposition from those who see them as a threat. But it has also provoked an equally strong determination in their supporters to liberate us from King Oil and the dangers of nuclear waste. As with the earlier personal computer revolution, inventors are making breakthroughs in home workshops and garages as well as in professionally run laboratories. Observers of these developments predict that this revolution will have more of an impact than PCs have had. These inventions could do more than transform our homes, vehicles, and factories; they could also help clean up the water, air, and soil.

Why haven't you heard about the push to develop radically different energy technologies before? Scrutiny of the new-energy field reveals a complex picture, with dark areas of greed, corporate lobbying, international energy politics, bureaucratic inertia, academic resistance, secrecy, and inventors' paranoia. However, the bright spots of irrepressible new discoveries are growing ever faster and are showing up in unexpected places.

What is space energy? We deal with this subject more fully in Chapter 4, but let's start with a brief explanation. For most of the twentieth century, science thought of space as being empty. It is not. Space—both interplanetary space and earthly space—is incredibly dense with energy, a sea of energy. This sea of energy fills everything, including our own bodies. Therefore, we can't sense it, nor can we measure it against something else. But there are inventors who say that they have been able to harness this energy, to pull it out of the air and put it to work, without pollution or fear of scarcity.

As wonderful as it sounds, space energy is not our only new-energy option. There's cold fusion, a nuclear reaction that can take place on a tabletop. There's hydrogen, a clean fuel that can be extracted from water. There's heat technology, which turns waste heat into electrical power. There's low-impact hydropower, which can tap the energy of our rivers and oceans without dams and flooded valleys. And there's other new energy possibilities.

Futurist John L. Peterson, in a report prepared for the United States Coast Guard, calls space energy a major force for change. He says that once the technology is improved and developed into marketable products, "all existing energy production methods become obsolete." And he doesn't see this happening thirty, or twenty, or even ten years from now. He sees it happening soon.

What does this mean? A switch from a world economy based on fossil fuels to one based on abundant, clean new energy would dwarf any other event of our times. Politicians would be disoriented as they move from the familiar oil-war mentality to an unfamiliar situation in which there would be abundant power for all. In comparison to the crumbling of the fossil-fuel worldview, the fall of the Berlin Wall would be a small blip in history.

Inventors in this field have often been individuals without advanced scientific training, working in small workshops. The standard scientific viewpoint has been that these inventors didn't know what they were doing, that these new energy sources cannot exist because they go against the known laws of physics. In recent years, however, some highly trained scientists have defied that viewpoint and started taking new energy seriously. Around the world, respected physicists are recognizing that official science has painted itself into a corner. For too long, orthodoxy ignored mounting evidence in support of new energy. Now it seems as though the laws of physics will have to be interpreted in a new way.

I believe we are in a new-energy breakthrough period, with inventors developing revolutionary energy devices that could power ships, homes, aircraft, greenhouses, and industries. This power can also be used to desalinate seawater, irrigate deserts, and help fuel a massive environmental cleanup.

To picture some changes these new Galileos expect to bring about, imagine yourself buying an enhanced energy converter—smaller than, say, a portable piano keyboard. This fuel-less device contains no moving parts, yet it puts out enough power to run your home or your new electric vehicle without being plugged into a wall socket or a battery. Since you no longer have to pay a utility bill or buy gasoline, you have the money to lease or purchase the converter. After the hardware is paid for, the electricity you use is free. You can live anywhere, from a mountaintop to a houseboat, because you can heat and power your home cheaply.

When can you buy a new-energy device? That depends on factors discussed in this book. A lot of new-energy hardware is in the crude premanufacturing stage—where the aerospace industry was in 1903 when the Wright brothers flew their homemade aircraft for less than a minute along a beach. However, a team effort and some substantial investment could bring some of these inventions to store shelves soon. Japan and a few other countries without oil wells—countries strongly motivated to find new sources of energy—show the most interest in such a team effort.

Is harnessing the energy of space an impossible dream and are its proponents merely kooks, as some new-energy debunkers would have you believe? As a skeptical journalist, I expected for years to find that, yes, the guardians of official science were right and that it is impossible to run machines on water, much less on energy from thin air. My expectations were reinforced when I looked at some of the amateur literature of what is called "fringe science," written by people whose ideas are all too often swept into a corner and labelled as "crazy."

During most of the 1980s, I was still uncertain whether the claims of the new-energy inventors could possibly be true. My degree is in sociology instead of the physical sciences, and my work experience is mainly in journalism. I, too, had absorbed the everyone-knows attitude that such inventors' claims violate the laws of physics, and are therefore ridiculous. That unquestioning attitude began to change slightly in 1982 when I met an inventor of an unorthodox energy machine, and during the rest of the decade I began a search for answers.

Is so-called free energy possible? Increasingly, it did look as though it was possible to convert a previously unrecognized energy source into useable power. I travelled, photographed, and interviewed, but as a journalist I was trained to remain skeptical and expect that the mavericks were mistaken. However, the weight of evidence pointed to the reality of useable new-energy inventions.

The aim of this book is introduce you to this fascinating world, the implications of which should be discussed publicly. Issues raised by the prospect of cheap electrical power and decentralized sources of abundant, clean energy are crucial to the economies of countries, and to the well-being of individuals. This book is intended to be a discussion-starter.

To make it easier to see the emerging new-energy picture, this book is divided into several parts. After Chapter 1, which discusses new-energy basics, Part I looks at the history of new energy, introducing people active in the past who were ahead of their times. Part II takes a closer look at space energy, at the physics behind it and at some of the inventors who have captured it. Part III explores the other new energy technologies mentioned earlier, such as cold fusion and heat technology. And Part IV looks at the problems and benefits involved in the development of new sources of energy.

The inventors you will meet in this book represent only a small portion of the new-energy scene. While this book champions the lone inventors and mavericks, I do not mean to underestimate contributions from the worlds of academia, government, and business. These institutions, though, are backed by well-financed public relations efforts. This book is intended to balance the picture.

I want to relate the stories of these science renegades to not only explain new-energy theories and devices, but to also show the harassment these inventors have encountered. My aim is not to arouse an "Ain't it awful" reaction. Instead, I wish to draw public attention to the situation, in the hope that public understanding will smooth the path of these energy visionaries. We all have a stake in their success.

Even now, the suppression is clearing away, as if the winds of change are blowing through the smog of our past ignorance. Many brilliant minds around the world are making breakthroughs in revolutionary energy technologies by using a variety of approaches. It is a coming energy revolution.

One researcher—Brian O'Leary, Ph.D.—gave up lucrative employment for the challenging life of an author and independent scientist. This frontier scientist knows the academic world, having been on the faculties of the California Institute of Technology, Cornell University, and Princeton University, and having published more than 100 scientific papers. He is familiar with politics, having been an energy consultant to Congress, and an energy advisor and speechwriter to presidential candidates. He has also worked with NASA on the Apollo program.

In 1991, he cofounded the International Association for New Science, and later helped start the Institute for New Energy. Thanks to the institute, I had the privilege of getting to know Brian O'Leary and his partner, artist Meredith Miller. I was honored when Brian O'Leary agreed to write the foreword for this book.

1
Quantum Leap

Shouldn't the government be paying attention to a field that has the potential for creating hundreds of thousands of new jobs at all skill levels?

—Eugene Mallove,
Infinite Energy magazine editor

Throughout the twentieth century, there have been individuals who have insisted that humanity could pull useable energy out of thin air. Their views have been unacceptable in academic circles. Some of them have been told that business interests did not want them around. Some of them have even been shot at or had their laboratories broken into.

Then, in 1986, these lone inventors found out that the United States Air Force wanted a firm to do research into ways to use "esoteric energies heretofore unknown, including the zero-point fluctuation dynamic of space." In other words, the Air Force was exploring the use of space energy, one of the new-energy sources examined in this book. However, the home-laboratory tinkerers decided that such a request was not meant for them. No more was heard about it.

Why is this research unknown to the general public, and why aren't more scientists working on it? As one physicist says, "Amazing effects are being discovered by people outside the scientific mainstream. Unfortunately, most scientists are missing the chance of a lifetime."

SCORN, THREATS, AND THE MISLEADING LABEL
OF "PERPETUAL MOTION"

It seems that the cards have been stacked against the independent inventor of new-energy devices. The public did not hear about the military's interest in "esoteric energies heretofore unknown," and thus people have only snickered at the lone inventor down the street who talked about "free energy."

Scoffing neighbors have been the least of the inventor's problems, as you will see in this book. Frequently, newspaper accounts would describe the inventor as an eccentric or a "perpetual-motion crank," without seriously evaluating the inventor's claims.

Those who have believed their wealth and power to be threatened by a possible energy revolution have been more sinister in their attentions. The harassments endured by inventors have included threats, destroyed equipment, and shootings. Patents have been denied, and plans for various devices have mysteriously vanished. Some inventors have died financially impoverished and spiritually broken.

One reason these inventors have had such a hard time of it has been the mistaken association between "free" energy and perpetual motion. Perpetual motion—generally in the form of a machine that, once set in motion, will go on forever without receiving energy from outside of itself—is impossible. Perpetual-motion machines of various designs have been deflating hopes and fleecing investors since the Middle Ages.

However, a new generation of energy-hardware breakthroughs has nothing to do with perpetual motion. The new-energy researchers have argued that they are harnessing the energy present in the vastness of space. Such devices could be seen as operating in an open system, not the closed system assumed in a perpetual-motion system. A closed system can be pictured as being in a closed box that contains only the machine and its fuel. The amount of fuel going into the machine is known and finite. On the other hand, the source of power in an open system is not limited to what is known to be in the box. The box is open, letting in an infinite amount of energy.

Has anyone been listening to the new-energy researchers? In the 1980s, their voices—from laboratories around the world—had not yet become a chorus. At gatherings of no more than a few hundred people at a time, they showed their wares and were inspired by fellow visionaries. Now, their conferences still attract only hundreds, but thousands of other people are dialing into new-energy computer bulletin boards and reading new-energy magazines.

Before we can discuss new energy, we have to ask ourselves some basic questions: What is energy? And what role has energy played in human history?

WHAT IS ENERGY?

Energy is the ability to do work. Traditionally, all energy has come, directly or indirectly, from the sun. The only sources of energy in current use that do not come from the sun are tidal power, which comes from the gravitational pull of the moon, and nuclear energy. (See "How Energy Is Measured and Produced," page 4.)

The Sun as Energy

The sun is believed to be a large nuclear fusion reactor, combining hydrogen atoms into helium atoms at a rate of four hydrogen atoms to one helium atom. This fusion releases the energy that we receive as heat and light.

Light reaches the earth as photons, those units of radiant light that enliven plants through a process called photosynthesis, in which the plants turn the energy of the sun into food energy. Animals convert this food energy into muscle energy, either directly by eating the plants or indirectly by eating plant-eating animals. Humans have used both their own muscle energy and that of domesticated animals, such as oxen and cattle, since the dawn of recorded time.

Some of the sun's energy is stored by trees in the form of wood. Humans developed this source of energy when they discovered how to make fire.

The sun also creates wind by constantly warming certain areas of the atmosphere more than others, and thus causing the air to move in different masses. This energy has been used to turn windmills. The movement of air masses also helps to cause water droplets in the air to form clouds, which in turn produce rain, snow, and other forms of water. This water gathers in streams and rivers, and flows to the oceans. Along the way, it often forms waterfalls. The energy of falling water has been captured, first by simple waterwheels and later by turbines running electric generators.

How Energy Is Measured and Produced

The word "energy" comes from the Greek words "en," meaning "in," and "ergon," meaning "work." It is a general term that covers all sources of heat and power. The unit used to measure energy is the joule. When used to measure mechanical energy, it represents the amount of work that is done when a force is applied to move an object with a mass of 1 kilogram, or about two pounds, a distance of 1 meter, or about a yard, in 1 second.

In terms of electric energy, a joule represents the work done in 1 second by 1 ampere flowing through 1 ohm, while a volt measures the electric potential that exists when 1 ampere flows through 1 ohm. Amperes are used to measure the amount of electric current flowing through a system, while ohms are used to measure how much the system resists the flowing current. A watt measures the work done by a current of 1 ampere under the pressure of 1 volt.

Fuel is energy in an undeveloped form. Coal, oil, and gas contain chemical energy, which is released when they are burned. Certain unstable elements, such as certain forms of uranium, are good sources of nuclear energy, which is released when the atoms that make up these elements are either broken apart or smashed together.

Electricity is generated by using the energy released by these fuels to boil water into steam. The steam is used to turn a rotating turbine, which in turn runs an electric generator.

Chemical energy can also be used in an internal combustion engine, such as those found in cars and trucks. In these engines, the fuel is burned in cylinders. This produces energy that drives pistons, and the up-and-down motion of the pistons is then converted into a rotary motion that drives the wheels.

Energy density measures the amount of energy available in a given amount of fuel. For example, hydrogen has nearly three times the energy density of oil.

Fossil Fuels and the Industrial Age

The human use of energy was relatively modest until the discovery and exploitation of fossil fuels. The sun was also responsible for this form of energy. Through photosynthesis, photons energized the giant ferns and dinosaurs that died in prehistoric swamps. The carbon molecules in the bodies of these plants and animals were eventually squished into different forms—coal or oil—after the earth's crust buckled, tucking the swamps deep underground. As the fossils decayed, natural gas filled caverns inside the earth.

The first fossil fuel to be widely used was coal, especially as used in the steam engine, which pulled humanity into the Industrial Age. In Britain, James Watt developed the steam engine into a modern form between 1763 and 1787. By 1850, coal-fired steam engines powered railroad cars, and steam also took to the streets in the form of wheeled vehicles powered by steam engines. Steam was put to its most efficient use on the water, where steamboats eventually overtook even the fastest sailing ships.

The nineteenth century also saw the development of oil as an energy source. This, in turn, allowed the development of other forms of transportation in the twentieth century, when automobile use became widespread and the airplane was invented. Before the century's end, men walked on the surface of the moon and returned to tell of it, and their voyages were powered by fossil fuels.

It was found that fossil fuels were excellent sources of heat for both commercial buildings and dwellings. It was also found that these fuels could be used to run turbines that could, in turn, run electric generators. Electric power—human-generated lightning—was soon used for an ever-increasing number of uses, from trolley cars to street and house lights to industrial motors. This further quickened the pace of industrialization.

While things moved faster, the source of energy remained the same—fossil fuels. One science writer says that "every day, the human race consumes a million billion kilojoules of energy by burning fuel," or the amount of energy contained in eight billion tons of oil. The poisons smoking out of that massive daily burn, such as sulfur and nitrogen byproducts, are well known, as are their effects on humans, including cancer, birth defects, and a host of other physical problems.

Pollution is not the only problem associated with our use of fossil fuels. Another problem will be the increasing scarcity of fuel as the world's reserves are used up. One magazine article states, "Our oil supply formed over units of geologic time—millenia, epochs, eons—but it is being used in units of human time—centuries, decades, years." According to one estimate, 950 billion barrels of recoverable oil exist in discovered fields worldwide. Undiscovered resources may hold about 500 billion more barrels. And so far, the world has produced and consumed more than 650 billion barrels of oil.

If we continue to consume oil at the present rate, the geologists estimate that the supply would last seventy years. However, people in developing countries want to obtain higher standards of living, which means more energy consumption. And neither natural gas nor coal represent a long-term solution. There is an estimated forty-year supply of natural gas, and coal, the most plentiful fossil fuel, produces the most pollution.

Another result of our hunger for energy has been the development of an economy built upon fossil fuels. The production, transportation, and use of these fuels requires a large, complex system—everything from oil refineries to power plants to the gas stove in your kitchen. This means that a lot of the world's financial resources are strongly tied to the fossil-fuel economy.

Nuclear Energy: No Fuel Like a New Fuel

Humanity has tried one new source of fuel on a megawatt scale—splitting atoms to release the energy within them. It started in 1942, when the first reactor was built at the University of Chicago. Three years later, the first atomic bomb exploded in New Mexico. A few weeks thereafter, atomic bombs were dropped on Hiroshima and Nagasaki, Japan. After a promise to use "atoms for peace," nuclear engineers created the first nuclear power plant, which started up in southern Idaho in 1951.

The dangers posed by radiation are well known: radiation sickness, higher rates of cancer and reproductive problems, long-term environmental contamination. The world learned exactly how dangerous nuclear power could be in 1986, when the Chernobyl plant in the Ukraine experienced an explosion and fire that released a cloud of radioactive gas. Downwind from Chernobyl, in the country of Belarus, less than 10 percent of the children in a total population of 10 million people are considered healthy. The rest suffer from a number of illnesses. For example, some types of cancer are occurring at rates more than 120 times their pre-accident levels.

Nuclear power has also become expensive. What was advertised as Power Too Cheap to Meter has turned out quite differently. The cost of building a nuclear plant has escalated, spurred by inflation and the costs of safety and testing. In 1980, it cost $1,135 per kilowatt to build a plant. In 1989, it cost $4,590 per kilowatt—about four times as much. Some reactors have cost more than five times their original estimates. Costs of more than $3 billion a plant are common.

This cost is being passed on to future generations. A typical nuclear power plant creates more than thirty metric tons a year of spent fuel, much of it in the form of highly radioactive waste that will pose a danger to life for thousands of years. About 20,000 tons of uranium waste are stored in pools at reactor sites in the United States. And despite spending hundreds of millions of dollars to study the problem, officials at the United States Department of Energy are still scratching their heads over long-term storage of this material. Whatever is done, it will cost hundreds of millions of dollars more to build repositories, transport the waste, and watch over it until it becomes harmless.

Humanity has since had its doubts about nuclear energy. For example, France is having second thoughts about its decision to become the Nuke Country. To the surprise of many, industrialists who want to generate their own power have joined environmentalists in attacking the ambitious French nuclear program. Why? According to one writer, the government has undermined the country's economy in order to bail out its nuclear industry. For example, in its first twenty-five years of existence, the state-controlled Electicité de France—France's main power supplier—has accumulated debts of 230 billion francs, and is facing an unknown bill for nuclear waste disposal and reactor decommissioning.

The other form of nuclear energy that has been studied—hot fusion, in which energy is created by the coming together of atoms—has also been a heavy expense to taxpayers. In 1950, Soviet physicists figured out how to make a *tokomak*, a huge, doughnut-shaped device used to create controlled nuclear fusion, such as that which is believed to occur within the sun. The United States began its own research program in 1951. In Princeton, New Jersey, thirty-five years later, a tokomak fusion test reactor achieved temperatures of 200 million degrees Kelvin—or almost 36 million degrees Fahrenheit. However, in April 1994, the head of a Congressional energy subcommittee said that after nearly $10 billion of taxpayer-funded research, the Department of Energy was still decades away from useful fusion energy. One legislator said, "A tokomak reactor is unlikely to be a commercially feasible power system."

THE NEXT STEP: NEW ENERGY

It is becoming obvious to many people that the fossil-fuel line must come to a dead end, and that alternative sources of energy—renewable, nonpolluting energy—must be found. A number of the alternatives that have been investigated, such as solar, wind, and tidal energy, are fairly well known to the general public.

However, this book deals with sources of energy that are not that well known. One is space energy, also known as "free energy" or "zero-point energy." Science has known for a long time that heat causes movement in the molecules that make up all matter. But even at the absolute lowest temperature, or zero-point—the point at which molecules cease to move—the atoms that make up the molecules still jitter with fluctuations of electricity. Thus, all space everywhere—including earth—is filled with energy.

The electricity we use every day moves in definite paths through wires. However, the movements of the hidden electricity of space are random. Therefore, scientists have traditionally believed that this invisible, seething motion could not be harnessed to do useful work, in accordance with a law of physics that says, "Whatever is random must forever remain random." However, since space energy fills everything, including our own bodies, scientists can neither sense it nor measure it against something else. The challenge is akin to that of measuring a half-cup of water when the cup is held underwater in a lake. Today's standard science lacks instruments refined enough to detect this energy.

Nevertheless, there have been individuals who say, in defiance of standard physics, that they have been able to build devices powered by the background energy of space. The inventors say that these experimental models draw energy out of this unlimited source and, when perfected, could run indefinitely.

In Part I, we will meet yesterday's new-energy pioneers. In Part II, we will first take a closer look at space energy, and then we will meet inventors who are working with today's space-energy technologies:

• *Solid-State Devices.* In Chapter 5, we will meet three men who are working on solid-state space-energy devices—devices that don't use any moving parts. Two of these inventors are on the cutting edge of high-tech science, while the other has gone for a low-tech approach—energy from ordinary rocks.

• *Solid-State Magnets.* In Chapter 6, we will meet a man who used the deceptively simple power of magnetism to draw energy from space in a solid-state device, and the inventors who are attempting to follow in his footsteps.

• *Rotating-Magnet Devices.* In Chapter 7, we will meet inventors— from the United States to Japan to India—who have put magnets in motion to draw on space energy.

Space energy is not the only new energy technology that researchers are working on. (See "Spirals of Energy," page 13, for a link between space energy and other sources of new energy.) In Part III, we will explore a host of other amazing possibilities:

• *Cold Fusion.* In Chapter 8, we will meet inventors who say that the effort science has made so far to create nuclear hot fusion—an expensive and technically difficult process—is unnecessary. They say they can create cold fusion in jars that sit on a tabletop.

• *Hydrogen.* In Chapter 9, we will meet two inventors who have learned to harness hydrogen—one of the universe's most plentiful substances—using technology that could allow people to drive past the gas pump.

• *Heat Technology.* In Chapter 10, we will meet inventors who say they can turn the waste heat produced by many common processes into cheap, clean electricity.

• *Low-Impact Water Power.* In Chapter 11, we will meet inventors who have found ways to use one of the world's oldest energy sources—the power of moving water—without the need for expensive, environmentally troublesome dams.

• *Other Energy Technologies.* In Chapter 12, we will meet other inventors and visionaries, including a commune that harbors a new-energy device hidden away from the world.

INNOVATION AND SUPPRESSION

This book will introduce you to many of the innovators in the new-energy field. They strongly feel that it would be of priceless benefit to humanity if we could tap into the sea of energy. However, rapid development of such radically new sources of energy would turn the world economy upside-down. And those that profit from our present reliance on fossil fuels will not yield their profits—or their power—so easily.

The Innovators: Mavericks and Renegades

Those who have dedicated themselves to the new-energy field are a varied lot, from home-workshop tinkerers to highly educated scientists. Some work or have worked at the highest government and corporate levels, while others hide themselves away in the countryside. But what they all have in common is a conviction that there is a better way to power machinery and heat buildings than by burning fossil fuels.

Many inventors are driven by concern for their children and grandchildren; they want to pass on a cleaner, more health-enhancing planet. Others, who may or may not be idealistic, see the money to be made by securing positions in this new field, expecting that a multimillion-dollar energy market will be revolutionized.

Inventors have been tackling the challenge of new energy for more than a century, but until recently they lacked sophisticated electronics, space-age metals, and powerful magnets, and have not had instant access to advice through computer modems and faxes. In turn, skeptics have been tackling inventors' ideas through the years, but recently some skeptics have been won over by seeing some of the many new-energy machines or devices in existence.

Inventors and other researchers meet, in groups of several hundred to a thousand at a time, at conferences throughout the world. These conferences are sponsored by a dozen new-energy associations and institutes, including the Planetary Association for Clean Energy, based in Ottawa, Ontario, and the Institute for New Energy, based in Salt Lake City, Utah.

Salt Lake City is also the home of the journal *New Energy News*, which has compiled a database listing more than 1,500 new-energy papers and references to inventors' works. In North America alone, at least 15,000 people are keenly interested in this field, judging by circulation figures for publications such as *New Energy News* and *Extraordinary Science*, based in Colorado Springs, Colorado. And the use of computer bulletin boards is bringing an increasing number of inventors and new-energy researchers into contact with each other.

A few inventors of new-energy hardware are close to breaking through into major-industry status. For example, Clean Energy Technologies, Inc. of Dallas, Texas, plans to produce a cold-fusion system that puts out ten times more power—in the form of heat— than it takes in. And in Japan, a space-energy system is approaching the marketplace.

The Forces of Suppression

Human beings in general would benefit from the coming energy revolution, but many individuals would not. The forces of opposition include those who control the fossil fuels now used to power the world's machinery, and those in the military who see the sea of energy as a source of new weapons. As you will see in later chapters, many of the energy innovators claim to have been harassed by those who profit from the current system.

New energy has also been opposed by many in the scientific community. Throughout history, major upheavals in scientific thought have caught the established leaders in science unprepared. For example, in the early seventeenth century, those who believed that the sun revolved about the earth sent Italian astronomer Galileo Galilei to prison as a heretic for saying that the earth revolved about the sun. Such new beliefs upset the existing worldview, and are often deeply unsettling to those who adhere to that worldview.

Thus, opposition from the business, government, and science establishments has often resulted in attempts to slow down or suppress innovation. One space-energy researcher, a physicist, relates the following example from history.

Before strong lead glass was invented, only the wealthiest members of the French aristocracy could afford plate glass windows. Then, shortly after the French Revolution, lead glass became commercially available, coinciding with a flurry of home construction or renovation by members of the newly powerful middle class.

To the dismay of the candlemakers' guild, the large windows that were being installed had a bad effect on business. All those dark homes had meant steady customers for candles. But now, picture windows were opening up these homes to more hours of daylight each day. Candlemakers noticed a drop in demand for their product.

The candlemakers demanded that their new government pass a law to tax French homeowners—a stiff annual tax on each window wider than a few feet. The tax was levied.

How did the candle lobby get away with taxing sunlight? As the physicist explains:

> The guild argued that large windows were an aristocratic arti-
> fact; that they make homes too cold in winter, too hot in sum-
> mer; glass is fragile and unsafe; sunlight is bad for your
> health; large windows invite accidents, diseases, thievery and
> frivolity. . . . What the candle-manufacturers' guild actually
> did was not too different from what various special interest
> groups are doing today: preventing new clean, cheap, natural
> sources of energy from replacing the expensive traditional
> unclean, obsolete sources of energy.

In Part IV, we will take a closer look at the harassment endured by new-energy researchers, and at how the new-energy revolution will affect our world.

A QUANTUM LEAP INTO THE FUTURE

Until recently, the world's energy production system has been highly centralized, which has allowed it to be controlled by those who profit from it. Even the alternative sources of energy that have been most discussed in public assume that the centralized, tightly controlled nature of the system would not change. For example, the solar energy plans presented by most North American proponents are on a megaproject scale. Solar panels would cover many square miles of desert, where solar energy would break water up into its two elements—hydrogen and oxygen. The hydrogen would be then trucked or piped to customers, as natural gas and gasoline now are.

In contrast, most of the researchers you will meet in this book are convinced that the coming energy revolution is about freedom from complicated energy-delivery systems, about getting away from hundreds-of-miles-long pipelines and power lines that could be damaged by earthquakes or sabotage. It is about avoiding situations that leave average people literally power-less. In short, it is about decentralized clean-energy sources—power for individuals, families, neighborhoods, or businesses. One goal of new-energy researchers is to see thriving industries that can plan for the future without depending on foreign oil. This goal is opposed by a well-funded army of fossil-fuel lobbyists.

Spirals of Energy

What do space energy and other forms of new energy have in common? Many researchers think the common link is the three-dimensional spiral, or vortex.

Whirlpools and tornadoes are both types of vortices. They create a funnel of energy from top to bottom. To picture what a vortex looks like, picture the sort of spiral coil that you might find in a mattress.

Side view

When viewed from the top, the spirals in a vortex get wider and wider as you go from the inside to the outside. One can see this shape everywhere in nature, from seashells to fern fronds.

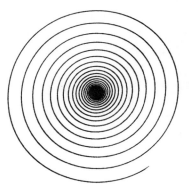

Top view

Motion along such a spiral can be either outward or inward. In outward motion, matter and energy are dissipated. In inward motion, however, matter and energy are created. Researchers believe that space energy and other forms of new energy follow an inward-spiralling path, and are in constant motion.

Another goal of some of the new-energy innovators is to keep the United States from falling behind economically. In response to the announcement of a breakthrough discovery in Japan, one researcher said:

> The U.S. government, once it wakes from its lethargy . . . will be faced with another Sputnik. . . . The financial community, once it realizes that overunity [more power output than input] devices have broken through legitimately, will also be in an uproar . . . they have an awful lot of capital invested in things that are shortly going to become very much less relevant. . . . Sputnik did not threaten to dominate all our industries, labor forces. This does. . . . The Japanese know that overunity energy systems are essential to national survival. They are already in the game like gangbusters. They've scored the first goal. When is the U.S. going to at least start shuffling its feet and preparing to run?

What is required, even more than research funds and material, is a willingness to look beyond the limits of what is now thought to be possible, a willingness to take a quantum leap forward into the next era. One publisher in the conventional energy field was present at a demonstration of the advantages of fiber optics over copper wire. Afterwards, he thought about electricity, and wondered: where is the equivalent leap forward, major change, or radically different thing? Even nuclear power uses heat from its atom-splitting to boil water into steam, and the electricity is made by the usual steam turbines. "We are still boiling water!" he says. "We are still using twenty-first century technology on top of nineteenth century plumbing!"

In the first part of this book, you will meet pioneering energy heroes who worked against great odds to develop unheard-of energy inventions.

Part I
Past Revolutionaries

New-energy technologies have a hidden history full of human triumphs, struggles, and setbacks. When they managed to build devices that seemed to defy the known laws of physics, certain individuals had their moments of fame, but their accomplishments never appeared in school textbooks. However, although most of the stories surrounding new-energy inventions have been unknown to the public, that may change. For example, Fox Television has run a program on the most well-known of the early new-energy inventors, Nikola Tesla.

Someone asked me why I wanted to look backward in this book, since we have geniuses making exciting advances in the new-energy field today. My reply is that some gifted individuals were so far ahead of their times that some modern-day scientists say they are still learning from those past revolutionaries, that there are still clues to be found in past research about how energy works.

There are also lessons to be learned by those of us who are not scientists. We can read about these past inventors both to appreciate their struggles and to see that "new" energy does indeed have a history.

2
Nikola Tesla — The Father of Free Energy

Ere many generations pass, our machinery will be driven by a power obtainable at any point in the universe. . . . Throughout space there is energy.

—Nikola Tesla,
Inventor

Dr. Nikola Tesla was once one of the most famous people on the planet. Today he is written out of our scientific and educational textbooks. What did he discover that caused his fall from grace?

—*Nexus* magazine

In the late nineteenth century, no one was more celebrated by the cream of New York society than inventor Nikola Tesla. Tesla, a Serb who had immigrated to the United States, often held court at his laboratory, where friends such as Samuel Clemens—better known as Mark Twain—posed for the first photographs ever taken by gas-filled tube lights. They stared open-mouthed at the middle of the room, where long sparks roared from Tesla's special high-frequency transformer. At times, their host would stand in the spray of high-frequency electricity, while a glass tube in his hand would light up with no wires attached. And in hotel dining rooms and private parlors, Tesla's creativity and intellect attracted other cultural stars, including author Rudyard Kipling, architect Stanford White, pianist Ignace Paderewski, and naturalist John Muir.

Tesla was a man of contradictions, cool and detached yet charming. Although a loner, he was also a showman. Slim and tall, always perfectly dressed, he commanded attention with his aristocratic posture and grace. His most striking feature was his magnetism—a combination of dark good looks, intense blue eyes, and an aura of mystery. It seemed that the world was destined to be his.

By the time Nikola Tesla died at the age of eighty-six in 1943, his inventions and theories had been largely forgotten and discredited, his plan to provide free energy worldwide discarded. Many of the later energy innovators who admired Tesla would also run into the problems of financial hardship and powerful opposition that helped to cause his downfall.

TESLA CLASHES WITH EDISON

Thomas Edison first met Tesla in 1884. Edison was already a wealthy, powerful man. Tesla was a new immigrant with little more than twenty dollars and a letter of introduction from one of Tesla's bosses at Continental Edison in Paris, where Tesla had been hired a few years earlier. Charles Batchelor had written to Edison, "I know two great men and you are one of them; the other is this young man."

Edison took Tesla on as an assistant. Tesla at first admired what Edison had accomplished by trial and error, and with only a grade school education. In return, Tesla won Edison's grudging respect by working eighteen-hour days, seven days a week, and by conquering difficult technical problems.

However, Edison soon lost his industrious new assistant. Tesla had described how he could improve the efficiency of Edison's generator, and Edison had clearly replied, "There's fifty thousand dollars in it for you if you can do it." But when, after months of work, Tesla did it and asked for his money, he was shocked to hear Edison say, "Tesla, you don't understand our American humor." Edison wouldn't pay. Tesla walked out.

Three years later, after time spent digging ditches on a New York street crew in order to earn a living, Tesla's luck changed. He got the chance to develop his system of alternating current (AC), for which he designed and patented a motor, generator, and transformer. Industrialist and inventor George Westinghouse of Pittsburgh bought all of Tesla's patents on the system, and signed a contract to pay Tesla start-up cash and stock, plus royalties of $2.50 per horsepower produced by the system.

Edison fought the development of AC. Edison's lamps ran on direct current (DC), which is a flow of electricity in one direction. DC can be transmitted for only a few miles on power lines. Tesla's alternating current flows back and forth in a regular rhythm. AC easily travels for hundreds of miles down wires—what we call high-voltage lines—with transformers at the receiving end to reduce the voltage for the customer's use.

Edison didn't want to hear about AC's advantages. He had a lot of money invested in a DC system, and saw AC as a threat to his business. His strategy in the War of the Currents included electrocuting dogs and publishing scare pamphlets, all in an attempt to link AC with death.

But despite Edison's efforts, Tesla and Westinghouse won. Westinghouse built an AC system for lighting the 1893 World's Fair in Chicago. Tesla was a star at the exhibition. In white tie and tails, and wearing cork-bottomed shoes for protection, he shared a stage with one of his Tesla coils—a device he invented that generated high-power currents. Its bolts of electricity crackled and snapped, and lit lightbulbs in Tesla's hands. The crowds loved the drama, and the exhibition's success led to the development of a hydroelectric project at Niagara Falls. Eventually, Tesla's distribution system delivered immense amounts of electrical power across the continent. Since the contract with Westinghouse gave Tesla $2.50 per horsepower, Tesla should have been assured a handsome income for life.

However, George Westinghouse was in financial trouble, with business competitors trying to squeeze him out of the power picture. Tesla remembered that Westinghouse had believed in him when others hadn't. And though Tesla certainly enjoyed money when he had it, it was more important to him to see Westinghouse's company survive. Therefore, Tesla tore up the contract, took a cash settlement, and walked away from the millions of future dollars assigned to him by the per-horsepower deal.

THE PROFIT MOTIVE VERSUS FREE ENERGY

While Tesla tore up a lucrative contract to help a friend, other men of his time were poised to grab as much money as possible. Tycoons prepared to make fortunes in utility companies. These men wanted the AC power system to cover the earth with power poles, transformers, and wires. Power companies would eventually dam rivers and encourage "better living through electricity."

Tesla, on the other hand, wanted to create a system of energy that would be transmitted worldwide for free. Tesla's proposed system was not "free energy" in the sense that term is used today—energy from an unlimited source—but in the sense that it would have been sent at no cost to the consumer. Unlike the War of the Currents, this was a war that Tesla could not win.

Tesla's Free-Energy Plans

Energy for anyone who sticks a tuned receiver into the ground? Yes, Tesla planned to transmit both messages—what we now know as radio—and energy without wires. This plan was radical enough to eventually cause Wall Street to slam doors in his face. It was a time when power monopolies would soon be floating in money; no one wanted to rock the boat. Corporate moguls such as banker J. Pierpont Morgan had already bought up copper mines. It did not take much insider information to figure out that transmission wires would someday cover much of the world with nets of copper strands.

As if oblivious to the schemes of monopolists, Tesla went on to propose a startling new idea—the worldwide transmission of free power. In 1893, the same year that he dazzled society by lighting the World's Fair, Tesla talked about earth resonance at the prestigious Franklin Institute in Philadelphia. Earth resonance was part of his vision for wireless power. It involved sending out electrical pulses of the correct frequency, or speed of vibration, through the earth to create waves of energy, just as a piano string will vibrate when another instrument at a distance hits the same note as the pitch to which the string is tuned. Some Tesla researchers also believe that he could have caused the air between the upper atmosphere and the ground to resonate like the air within a violin. This would also send out waves of energy. This energy would then be picked up by an antenna.

Such resonance would fulfill Tesla's dream, expressed in an 1897 speech, of "the transmission of power from station to station without the employment of any connecting wire." He saw a day when such a system would speed communications, control the weather, and supply limitless energy.

An ordinary person would have been too distracted by fame and the international lecture circuit to think about such things, but Tesla was not ordinary. His ideas and inventions were his obsessions, and in the next few years he wrote and was granted patents on processes for futuristic wireless transmission of power and messages, despite the chance that they might make his own previous inventions obsolete.

Wireless Power Is Tested

In 1899, Tesla relocated to the mountains of Colorado Springs to test his new ideas. In a high-altitude cow pasture, he built a high-voltage laboratory. It was simply a building wrapped around the world's largest Tesla coil, with a strange flagpole placed on top. There, in the shadow of Pikes Peak, he worked toward his new goal of sending electromagnetic vibrations throughout the earth.

Exactly what Tesla accomplished during his mountain sojourn is unclear. He kept sometimes-terse notes, retaining a lot of information on the device's operating principles in his head, and his notes have to be translated into today's electrical engineering terms. But Tesla legends feed on the facts of his Colorado Springs experiments. Like some god of lightning, he tuned up his massive transmitter coil, which was fifty-two feet in diameter, to create a twelve-million-volt discharge, and threw bolts of more than a hundred feet in length from the copper ball on top of his flagpole. The townsfolk kept their distance after rumors flew around that the famous inventor could make lightning which would kill a hundred people in one bolt. Meanwhile, thunder from the electrical discharges reverberated for at least fifteen miles.

Tesla returned to New York in January 1900, satisfied that he knew enough to carry out his vision of wireless transmission. He hired an architect to design a 154-foot-high wooden tower on top of a brick building on Long Island. It was to be used as a huge transmitter, with a mushroom-shaped copper electrode on the top. Tesla named the project Wardenclyffe, envisioning a station to broadcast power as well as communication channels of all radio wavelengths. The tower was nearly finished in 1902, along with the 100-feet square building that would contain a powerhouse and laboratory. But Wardenclyffe was never completed.

The Lightning Man is Struck Down

Tesla's vision of sending communications through "wireless intelligence" had convinced financiers such as Morgan to pay for the research, but they did not realize that he also intended to send free power to people everywhere. Tesla had left out that part when he spoke to Morgan, his main backer, in 1900 about financing for Wardenclyffe, concentrating instead on the opportunity Morgan would have to take monopolistic control over all radio broadcasting. But Morgan would give Tesla only limited funds. Three years later, desperate for money, Tesla admitted to Morgan his true intentions. While we'll never know Morgan's reaction to this news, the financier had investments in power-related industries, and in any case was not a man known for his willingness to give things away. He cut Tesla off.

Work continued sporadically as Tesla frantically tried to both find other backing and develop commercially viable products in order to pay the bills. Finally, in 1906, construction stopped, and eleven years later, after Tesla had lost his mortgage on Wardenclyffe, the tower was broken up for scrap metal.

A TRUE FREE-ENERGY GENERATOR?

There is evidence that Tesla was also interested in free energy in the modern sense of the term—energy converted into a useable form from a limitless source. In June 1902, an item appeared in *The New York Times* about a Canary Islander named Clemente Figueras, who claimed to have invented an electric generator that did not need a prime mover—that is, it didn't need an outside source of power. The day after the article appeared, Tesla wrote to his friend Robert Johnson, editor of *Century* magazine, that he had already invented such a device. And in 1934, Tesla was quoted in the *Times* as saying, "I expect to live to be able to set a machine in the middle of this room and move it by . . . the energy in the medium in motion around us."

Which of his many inventions was Tesla referring to? A scientist and historian from Utah, Oliver Nichelson, has examined the evidence. Nichelson says the one that at first seems to fit Tesla's description is the Apparatus for the Utilization of Radiant Energy, for which a patent was granted in 1901. Nichelson's research indicates Tesla was working on his "free energy" generator before he hammered out a major article for the June 1900 issue of *Century*, in which he describes sending power wirelessly. He writes that a device for getting energy directly from the sun would not be very profitable and therefore would not be the best solution. Researchers have read this to mean that Tesla had learned from his Wardenclyffe experience that a "free energy" device, such as his radiant energy apparatus, would never be allowed to reach the market, but that the tycoons would fund a profit-yielding wireless power system.

The *Century* article, however, focuses on a device that would pump energy from the surrounding air to light cities, instead of just being able to operate itself. For pumping energy from the cosmos, Nichelson has settled on Tesla's unusual Coil for Electromagnets, for which Tesla was granted Patent No. 512,340 in 1894, as "the most likely candidate." Nichelson explains that the coil's shape would allow the system to store tremendous amounts of energy while only using a tiny fraction of that energy to run itself. Nichelson compared it to a car that gets 100 miles a gallon with a very large, perpetually full gas tank.

THE FALL AND RISE OF NIKOLA TESLA

By the time Nikola Tesla died, his great achievements of the 1890s had been largely forgotten, and he was mostly remembered for his private eccentricities, such as his extreme fear of germs, or his willingness to invest an inordinate amount of affection in a pet pigeon and see reflected in her eyes a lifetime of hidden mystical yearnings.

Was Tesla's disappearance from the history books orchestrated by those who were threatened by his free-energy dreams? Some people believe that it was. University students are given the impression that he invented the Tesla coil and had a unit of measurement named after him, and that's about it. Tesla's name is not familiar to the general public. (See "Tesla's Champions Versus the Smithsonian" on page 25.)

If the power moguls did try to erase the public memory of Tesla's genius, the strategy did not totally succeed. Today, nearly every major bookstore on the continent harbors a Tesla biography. And since the late 1970s, there has been renewed interest in Tesla's research among inventors. Technical information on his theories and inventions is shared through fax machines and computer bulletin boards, and many of today's researchers regard Tesla as the father of the modern new-energy movement. They can empathize with the difficulties he endured in the face of powerful opposition.

Tesla is Spurned

I believe that the saga of Tesla's downward-spiralling finances centers on his monument to free broadcast power—Wardenclyffe. Margaret Cheney, in her classic biography, *Tesla: Man Out of Time*, writes about the complexities of Tesla's fall from fortune. Before Tesla's fall, she says, Tesla told an associate that J. P. Morgan once gave him a signed blank check and told him to fill in the amount he needed. After the fall, the banker would not reply to Tesla's letters, and the other financiers on Wall Street also turned their backs on Tesla for the remainder of his life. They may have thought him a dangerous dreamer—one of the comments Tesla made in a letter begging an associate for financial help was, "My enemies have been so successful in representing me as a poet and a visionary."

Other writers give different explanations for Tesla's descent. Science historian Stephen S. Hall supposes that the decline may have been a backlash from the academic community. Tesla did not play that game; he had no interest in submitting any article to an academic publication. Also, Hall thinks that Tesla's showmanship—his public displays, like the one at the 1893 World's Fair—may have caused "professional jealousy." Two other new-energy historians, Oliver Nichelson and Christopher Bird, say that Tesla was a mystery to his contemporaries: "So advanced were his concepts that science and industry of his day were unable to comprehend their essence and scope."

Dump Tesla, Hoist Edison?

Was a decision made earlier in this century to not only blackball Tesla financially but to also write him out of United States history, and install Edison as the official father of the electric age? I do not want to take away any credit due to Edison, who contributed heroically and prolifically to the age of electricity. However, I believe that the treatment of Edison, contrasted with that of Tesla, is part of a bigger picture—that of special interest groups attempting to manipulate public opinion.

The memory of Edison was hoisted onto a pedestal by a highly placed public relations effort. In 1929, more than fifty members of the military and industrial elite—including John D. Rockefeller, Jr., Julius Rosenwald, Henry Ford, Harvey S. Firestone, Herbert Hoover, and General John H. Pershing—formed a Committee for the

Tesla's Champions Versus the Smithsonian

A bust of Nikola Tesla.

While some of Tesla's modern supporters continue his research, others make sure that he is not forgotten by future generations. One teacher, John Wagner of Dexter, Michigan, contends that official history should remember Tesla for more than his eccentricities, which became more apparent as the inventor grew older. For ten years, until his retirement in 1993, Wagner taught about what Tesla accomplished in his heyday, instead of focusing on the man's declining years. Wagner made certain that his third-grade classes heard the complete story, including the fact that the Smithsonian, the country's national museum in Washington, had no permanent Tesla exhibit.

His students saw a double injustice, Wagner says—not only was there no Tesla exhibit in the Smithsonian, but Thomas Edison's large permanent exhibit displays one of Tesla's inventions, the polyphase generator. "Tesla's patent number is on it, but the public is left with the impression that Edison is responsible for it."

The children's indignation led to a campaign to "bust the Smithsonian"—the words on a T-shirt that Wagner's students sold. But when they offered to donate a Tesla bust to the Smithsonian, the curator of the museum's electrical division, Barney S. Finn, refused to accept the gift, saying, "We have no use for it."

In 1979, Finn and his staff wrote a small book entitled "Edison: Lighting a Revolution. " A seventeen-page section, entitled "The Beginning of the Electrical Age," purports to list all the people important to that beginning—even technicians who worked under Edison. But there is not a word about Tesla.

> *Wagner's students gained an unlikely ally in the rock band Tesla, which could be seen flashing pages from Tesla's patents on MTV. Wagner sent the band a letter, explaining the students' goal. In 1989, that letter resulted in a trip to Michigan by the California band, and twenty-eight excited children jostling onto the band's tour buses for a trip to the University of Michigan at Ann Arbor. There, in the engineering library, the children showed the musicians a bust —of Nikola Tesla as a proud young man—created with money raised by the previous year's class. The band agreed that the artwork should be cast in bronze, and offered to help in the children's efforts to get the statue into the Smithsonian.*
>
> *However, the museum continued to reject their efforts, and there is still virtually no acknowledgement of Nikola Tesla and his contributions in the Smithsonian.*

Centennial of Light to celebrate what was called "A World Wide Expression of Gratitude to Thomas Alva Edison on the 50th Anniversary of His Incandescent Lamp."

As part of this celebration, popular songwriter George M. Cohan wrote "Thomas A. Edison: Miracle Man," with lyrics such as "Oh say can you see, by the light he gives you and me./What a man he is, What a grand old `Wiz.'" The committee sent out a letter with Cohan's sheet music to community leaders and educators, saying that the song was a "tribute to the greatest living American. . . . You will aid in the tribute by renditions whenever appropriate."

Public emotion may have gone in a different direction if people had been told that Nikola Tesla had wanted to give them free access to electric power. But in contrast to the hymns of praise for Edison, Tesla was never celebrated by a committee such as the Committee for the Centennial of Light. And while some reference books concentrate on his work, others focus more on his personal attributes. For example, Isaac Asimov's *Biographical Encyclopedia of Science and Technology* sums up twenty-five years with the sentence, "The last quarter of his [Tesla's] life degenerated into wild eccentricity." (To which one inventor of today replies, "We should all be so wild.")

I believe that Edison is not the only inventor whose reputation has been pushed to the forefront at Tesla's expense. For example, why do textbooks ignore a United States Supreme Court decision in Tesla's favor over Guglielmo Marconi?

In 1901, when Marconi sent his famous radio signal across the Atlantic, Tesla said, "Let him continue. He is using seventeen of my patents." The Supreme Court set the record straight in 1943, after Tesla's death, ruling that Tesla was one of three turn-of-the-century inventors who had priority over Marconi in the patenting of radio-tuning circuits. School textbooks and other records of history, however, continue to elevate Marconi as the father of radio. A recent Smithsonian publication, the *Book of Inventions*, contains a section on radio. Tesla's work is not acknowledged, despite the Supreme Court decision.

Tesla is Rediscovered

The legend of Nikola Tesla refuses to die, despite the blackout of information in textbooks. A hundred years after his glory days, the new-energy press turns out many books for technical researchers on various aspects of his research, and an increasing number of young inventors are searching through his patents for clues. The researchers are scattered around the world.

Tesla's followers have gathered together in various groups. The largest is the International Tesla Society, based in Colorado Springs, Colorado, which sells books and videos, and runs a Tesla museum. That group's membership numbers more than 7,000. Tesla has also inspired a number of newsletters and magazines. (For more information, see the Resource List.)

The Russians have shown a great interest in Tesla's work. However, a lot of this research was done under Cold War conditions, and thus published information on it is scarce. For example, a top Soviet physicist, Nobel prize winner Peter Kapitsa, reportedly spent his final years in intense study of Tesla's writings. According to Cheney, Kapitsa wanted to contribute to Tesla's work on ball lightning—a part of his experiments on wireless energy transmission.

In the early 1970s, scientists from the former Soviet Union overwhelmed the staff of the Nikola Tesla Museum in Belgrade, Yugoslavia, when they arrived to examine Tesla's notes and devices. New-energy researcher Andrew Michrowski, Ph.D., of Ottawa learned about the exhaustive researches of the Academy of Sciences of the USSR when he visited the museum in 1975. The museum's director, Professor Aleksandar Marinčić, showed

Michrowski a thick book in small print. "Look what they found out. This was only the preliminary report," said Marinčić. Michrowski thinks that the Soviets may have experimented with very futuristic technologies as a result of their Tesla research.

Another Russian physicist, A. V. Chernetskii, Ph.D., inadvertently repeated a Tesla accident—Tesla's 1899 burning-out of the Colorado Springs power plant. In 1971, Chernetskii and a colleague performed an experiment that formed a large ball lightning and a storm of sparks. The surge of current ran along power lines at the Moscow Aviation Institute, overloading and destroying a power substation. They were trying to use Tesla's concepts to build a device that would produce more energy than it consumed.

There is continued interest in Tesla's wireless power distribution concept. It is a topic of discussion at new-energy conferences, and a number of groups, such as the Institute for New Energy, based in Salt Lake City, Utah, are following this research.

Other researchers are interested in Telsa's work on earth resonance. Tesla's followers look back with awe at his testing of powerful electromagnetic waves that would circle the earth and build up strength. Prominent experimenter Ron Kovac of Colorado has figured out that Tesla's equipment actually could generate very powerful waves for earth resonance, but says that today's experimenters are only beginning to understand Tesla's work.

Another Tesla invention that is being actively developed by today's researchers is his bladeless turbine. Turbines, engines that are moved by a current of air, water, or steam, are standard parts of conventional power-generation systems. But Tesla's turbine is more efficient, simple, and durable. It can recover additional energy from the waste outlet of a regular turbine, or it can recover other waste energy, such as that produced by oil or gas refineries.

One researcher says that automobile makers could use the bladeless turbine to eliminate the thousands of moving parts in a piston engine, which could double the engine's life. Jeff Hayes, founder of the Tesla Engine Builders' Association of Milwaukee, Wisconsin, says that in addition to the energy saved in building the car, the Tesla engine would triple fuel efficiency. He explains where the turbine fits into the concept of a super-efficient electric automobile: a bladeless Tesla turbine driving a high-frequency Tesla alternator driving an electric motor.

If there was no political opposition to marketing such a system, Hayes says, the technology could be developed "almost immediately." However, he feels that the government would not favor an engine that drove down fuel consumption, since a portion of government revenues comes from fuel taxes.

The Tesla turbine can also produce electricity when hooked up to generating equipment. One small company, Advanced Dynamics of Louisville, Kentucky, demonstrated such a setup at a 1995 new-energy conference. It generated enough electricity to light a bank of bulbs.

There were other trailblazers in the history of new-energy devices, with goals similar to Tesla's. In the next chapter, you will meet a sampling of these pioneers.

3
Other Innovators—
In Harmony With Nature

All these truly great applied physicists have learned to listen to the pulse-beat of nature, rather than the squeaking of chalk on a blackboard.

—Don Kelly,
New-energy researcher

Nikola Tesla has not been the only researcher over the past two centuries who has dreamed of providing abundant, clean energy to power our world. The six energy innovators we will meet in this chapter came from varied backgrounds. However, like Tesla, they believed in working with nature instead of against it. And like him, they ran into opposition, and the devices they produced were never used on a large scale for human betterment.

JOHN KEELY'S GOOD VIBRATIONS

Let us first consider a pioneer who predated Tesla. John Ernst Worrell Keely (1827–1898) of Philadelphia, a musician and carpenter, worked with sound and other forms of vibration to set machines into motion. He reportedly performed feats that twentieth century science is still unable to do.

For example, according to one of the stories discovered by Keely researcher Dale Pond of Nebraska, an apprentice spent six months with Keely learning how to build a motor.

"Are you ready to run it?" Keely asked after the final adjustments were made. "Then go ahead and turn it on."

The apprentice flipped the switch, but nothing happened. Keely walked over and put his hand on the fellow's shoulder, and the motor started.

A motor built to respond to a specific individual's touch? That was only one of the accomplishments attributed to Keely by writers of his day. According to historical documents, Keely performed other incredible feats:

• He built a machine that tunnelled through rock by dissolving the stone. His invention seemed to melt the rock as fast as he could move his device forward.

• He freed the energy within water in a manner similar to work being done by today's researchers, in which tiny bubbles are created in water by sound waves and energy is released when the bubbles implode. Observers saw an engine run on the energy freed by this device, which Keely called his Liberator.

Keely, musically sensitive and intuitive, discovered his effects by experimenting. His musical background allowed him to craft machines similar to musical instruments—as in the building of a violin, the machines were built to respond to tones in harmony, not conflict. But Keely's machines depended a great deal on what he called the "vibration tones" of the builder—the person's breathing and brainwave rhythms. It was as if a violin could only be played by the person who made it. Therefore, engineering his machines was not a simple matter of tuning electrical coils; they were much more sensitive than standard machines. Although he reached an advanced understanding of the science of vibrations, even Keely did not fully understand why his inventions worked. He also failed to develop machines that other people could operate.

According to Pond, Keely discovered more than forty of what he called fundamental laws of nature. Among other accomplishments attributed to Keely are the creation of frequencies at an extremely high range, as well as work in the fields of acoustical motors, ultrasonics, and control of extreme pressures and vacuums. Sometimes, it is hard to remember that this work was done in the nineteenth century.

Why isn't Keely's work well known today, and why isn't it studied in the scientific community? Part of the reason seems to be that Keely did not use scientific terminology to describe his work—he did not speak the language of science. Also, he was so far ahead of the science of his day that, like Tesla, his work was simply dismissed by many scientists.

Gentler Atomic Science

Conventional physics takes the sledgehammer approach—slam an atom with energy, break it, and see what is in it. This is the opposite of Keely's approach. He saw the atom as being like an orchestra, an assortment of vibrating parts producing various tones. This is now standard physics, but it was a revolutionary idea in Keely's day.

What really set Keely apart was his ability to act as the conductor of this atomic orchestra, his ability to get the atom to do his bidding. Pond says that Keely discovered how to bring two vibrations together so that they formed a third, different vibration. He uses the example of an opera singer shattering a wineglass with her voice, which represents the conventional sledgehammer approach. In contrast, Keely's approach would be to melt and reshape the wineglass by adding a tiny amount of energy to just the right tone.

Debunkers thrive on the Keely story, amusing their readers with the unusual names of his inventions—Disintegrator, Sympathetic Transmitter, Vibratory Accumulator, Tubular Resonator—and his extraordinary claims. There were also allegations of fraud. Newspapers of Keely's day considered the case closed when an investigator found a large metal sphere buried under the floor of his laboratory and slender pipes running throughout the walls after Keely's death. They declared that Keely had been using compressed air to power his experiments.

Researchers with some understanding of what he was doing, on the other hand, explain that was not the case. They say the sphere was not a hidden trick but a part of an experiment that was later stored under the floor. They also say that Keely used the pipes—too small in diameter to supply compressed air with any force—in advanced experimentation.

Keely and the Speculators

Keely's problems centered around the fact that speculators formed a company and pressured him for quick results so that they could make bundles of money. They squeezed him financially and psychologically. The history of the Keely Motor Company is a main reason why Keely was considered a fraud.

Keely began experimenting with vibrations and energy in the early 1870s. By 1874, he had gained some mastery over this force, which he called the "ether," but he had run out of money. Acquaintances offered to organize a company so that Keely would have the funds to develop an engine. The financiers of the Keely Motor Company expected immediate success.

However, years rolled by without Keely being able to produce a reliable motor, while his business partners manipulated and sold stocks. In 1879, the company faced bankruptcy. Keely consented to a complicated consolidation plan whereby he would assign two other inventions to the company in return for a fraction of the stock and a very small amount of cash.

Three years later, some stockholders sued Keely for not fulfilling the contract. One stockholder who was not involved in the suit wrote a letter to the *Philadelphia Evening Bulletin* in Keely's defense, saying that the money the stockholders had invested went not to Keely and his work, but to dishonest promoters within the company who had sold stocks and pocketed the proceeds. This view was echoed by Clara Bloomfield Moore, a wealthy widow who eventually became Keely's financial backer and biographer. She wrote that the plan was "prepared by schemers" and that "public statements that Mr. Keely has been supplied with large amounts of money from the company are untrue."

Despite this support, the Keely Motor Company fiasco sent the inventor to prison briefly in 1888. "Mr. Keely is his own worst enemy," wrote Moore. "When suspected of fraud, he acts as if he were a fraud." She was referring to an emotional outburst in which Keely destroyed instruments that it had taken him years to make. Moore said that Keely did this because he could not handle the insulting suspicions of arrogant scientists, but that his behavior resulted in "the suspicion that his instruments are but devices by which he cunningly deceives his patrons."

In 1890, a publication called *New York Truth* seemed to reflect the opinion of the time. "While Keely was hampered by mere tradesmen . . . more anxious for dividends than discoveries, he could do little save turn showman and exhibit partial control of the harmonies of nature."

Dale Pond says that modern science has vindicated Keely's work. Now that he is being taken more seriously, the Keely mystery deepens. What happened to the bulk of his writings? No one is sure. But researchers such as Pond are rebuilding Keely's machines and continuing his experiments.

WALTER RUSSELL AND THE INVISIBLE GEOMETRY OF SPACE

Walter Russell (1871–1963) was another energy researcher whose work is being reexamined. According to the Russell foundation archives, Nikola Tesla was so impressed by Walter Russell's theories that he advised Russell to lock up his knowledge in the Smithsonian for 1,000 years, until humanity was ready to use it wisely.

Russell was an acclaimed artist, musician, philosopher, and author. He taught himself science so successfully that he was awarded an honorary doctorate from the American Academy of Sciences. Russell was so far ahead of his time that in 1926 he predicted the existence and characteristics of tritium, deuterium, neptunium, plutonium, and other elements that weren't discovered until the 1930s and 1940s. As the president of the Society of Arts and Sciences in the 1920s, he had numerous occasions to express his views on matter and energy, which included an advanced understanding of what we now call space energy, the background energy of the universe described more fully in Chapter 4.

The Way to Cheap Hydrogen

Modern researchers have been able to confirm Russell's work. In the 1990s, three Colorado men—research chemist Ron Kovac, electrical engineer Toby Grotz, and naturopathic doctor Tim Binder—did in-depth laboratory research to see if Russell's theories and experiments hold up. They do. The trio repeated a Russell experiment from 1927—verified in that year by Westinghouse Laboratories—that demonstrated a cheap and efficient method of producing hydrogen. This would allow the development of a hydrogen-based fuel economy, discussed in Chapter 9, which would be virtually pollution-free and based on abundance instead of scarcity.

Space Energy and the Government

Russell also built a device, which he named the Russell Optical Dynamo Generator, that he claimed had captured space energy. Detective work by Toby Grotz turned up the original blueprints of this device, found in a Colorado basement.

The owner of the basement was an associate of a general with the North American Air Defense Command (NORAD)—the defense agency responsible for protecting North America from a nuclear attack. Russell worked on his device with both NORAD and scientists from the Raytheon Corporation. Grotz says that NORAD was interested in the generator because Russell claimed that it could not only produce more energy than it took in, but could also be used to create new types of extremely powerful radar.

In 1959, officers from NORAD in Colorado Springs visited Russell and his wife and research assistant, Lao, at their home in Virginia. They agreed that the Russells would report on their findings. On September 10, 1961, the couple reported that the Russell generator had worked, and that the president of the United States could announce to the world that a new, safe power source was available.

However, the Russells' conviction that they had demonstrated a way to convert space energy into electric power failed to capture the interest of anyone except NORAD, and there is no public record of what NORAD did with it. At the time, conventional science termed the discovery "nonscientific," and the public never heard of it.

Today's new-energy researchers are very interested in the theories behind the Russell generator. Russell said that the universe consisted of electric energy, and that nature multiplies power by concentrating this electricity—or space energy—until it forms matter, such as a star or a planet. Russell's device recreated this natural accumulation of power.

The Russells founded the University of Science and Philosophy in Waynesboro, Virigina, and researchers affiliated with the university are continuing his work. Grotz and his colleagues also want to pursue this research.

THOMAS HENRY MORAY:
BULLETS AND THE SEA OF ENERGY

Thomas Henry Moray (1892–1974) discovered the writings of Nikola Tesla in 1900, when Moray was a young boy with a home laboratory in Salt Lake City, Utah. One paragraph stuck in his mind, in which Tesla said that a form of energy pervaded the universe, and that if this energy was in motion then it could be used to generate power. The young Moray took it as a challenge.

Many people lose interest in those things that captivated them in their youth. But Moray, the son of a businessman, became an electrical engineer who pursued his dream—the idea that humans could mine energy from the cosmos by stimulating and amplifying existing oscillations in space.

Moray not only believed in this idea, he proved it publicly. His Radiant Energy Device worked for days at a time, converting space energy into useable power, and it was well documented and witnessed by respected authorities. Without any moving parts, the tabletop device produced a strange form of electricity that lit light-bulbs, heated a flat iron, and ran a motor.

But Moray's device was destroyed and his family suffered all kinds of harassment, apparently by those who did not want such a device to become available to the public. Now, his sons are trying to pick up where he left off.

Transistors and the Oscillations of the Cosmos

In 1939, Moray used a specially built apparatus that put out fifty kilowatts of useable electric power. According to one physicist, the experiments were witnessed by distinguished scientists. The fifty-five-pound Radiant Energy Device reliably pumped energy from the energetic ebb and flow of space. Moray's experiments indicated that energy comes to the earth in continuous surges, like the waves of the sea, and he channelled those waves of energy into the device. The device worked continuously for days without any sign of deterioration, according to reports by reliable witnesses.

The device included Moray's pioneering use of transistors, in which Moray was quite ahead of his time. During the first two centuries of research on electricity, scientists were interested in discovering insulating materials that would hold electricity captive, and conducting materials that would allow scientists to control electricity and experiment with it. They were not much interested in substances that were neither good insulators nor good conductors. However, these in-between substances, called semiconductors, were eventually found to have practical uses in that they allowed controlled voltage changes within electrical circuits.

In 1948, Bell Telephone Laboratories was credited with developing the first transistor—basically a sandwich made up of two or more types of semiconductors soldered to a base. The transistor allowed advanced electronic equipment, such as stereos and personal computers, to be built, because transistors are much more compact, durable, and long-lived than the old vacuum tubes they replaced.

Moray's supporters claim that he had invented the transistor almost ten years before Bell Labs. But whether their claims are accurate or not, Moray certainly should have gone down in the history books for his invention of the Radiant Energy Device, which held promise as a clean new-energy breakthrough.

Moray is Harassed

Having heard Moray's story from new-energy researchers, I interviewed his sons—John, a physicist who teaches school in Salt Lake City, and Richard, also a physicist, who lives on a ranch in Canada. Richard warily answers questions about his father. He explains why he zealously guards his family's privacy, and why he is not promoting Radiant Energy in the public eye: "I don't want to subject my family to what we were [subjected to]. . . . I saw my mother shot at. I saw my father shot at."

Richard will never forget that when he was a child the family had to order a bulletproof car, a car that did not stop their mother from being fired upon. In one incident, gunfire smoked from a mysterious black sedan and a shot slammed into the car as she drove the children around town. Fortunately, no one was hurt.

Their father was attacked while in his laboratory, resulting in a bullet in his leg. Henry Moray was an excellent shot with his pistol and could have killed the unknown assailant, Richard says, but his father was not a violent man.

Moray was harassed in other ways. His home and laboratory were broken into repeatedly when the family was not at home. In *The Sea of Energy*, a book he wrote about his father's work, John says that his mother received anonymous phone calls telling her that her husband's life was "'not worth a plugged nickel' unless he cooperated on Radiant Energy."

In the most serious—but poorly documented—incident, a man named Felix Frazer, who was working in Moray's lab, took a sledgehammer—or, as some reports say, an axe—and destroyed the Radiant Energy Device. The man's motivation may never be fully known. But it is known that he wiped out years of research and development, and damaged beyond repair some of the device's most vital components. John says that his father built another device later on, but then dismantled it, ostensibly for parts.

Who were Henry Moray's mysterious opponents? In the mindset of his times, Moray believed that the harassment and the destruction of his device was part of a communist plot. Other energy researchers think that the explanation is simpler, involving greed and perhaps even professional jealousy. The jealousy came from other scientists. The greed came from companies Moray worked with, which John and his father believed were mismanaged, and which disappeared with part of Moray's money.

Despite the death threats, Moray repeatedly showed his strange electric generator to creditable witnesses. The only threat that stopped him from further demonstrations came in the form of advice from his patent attorneys in Washington. They told him that under the patent laws, he could lose his rights to a patent if he continued to show his invention to everyone.

However, the United States Patent Office was not much help. The office rejected seven patent applications for the Radiant Energy Device because it did not fit the known physics of the time. Moray's semiconductor technology was so far ahead of its time that the patent examiner said he could not see how the device could work.

John and Richard have devoted much of their lives to finding the large amounts of money they believe are necessary to fund the device's expensive final premanufacturing stage, which would standardize its parts so that it could be produced in quantity. Some researchers believe that T. Henry Moray's secrets died with him and that his sons would not be able to replicate his device even if they had the multimillion-dollar funding required. Moray's sons, however, say they have inherited all his laboratory notes, and plan to continue his work.

THE MAGNETIC MOTOR OF LESTER HENDERSHOT

Is it possible that inventors have met premature deaths because of their inventions? That may have been the case with Lester J. Hendershot (1898–1961). His twenty-pound device converted power from the earth's magnetic field into enough electricity to run a television set and a sewing machine for hours at a time in his living room, according to his son and dozens of friends and associates.

Hendershot, from Elizabeth, Pennsylvania, held a number of jobs while he worked on his device, from fireman to mail truck driver to civil engineer. Ed Skilling of Illinois, an electronics engineer who worked with the inventor for a few years, says that when he met

Hendershot in 1958 he had expected to see a fast-talking con man out to take people's money. Instead, he found an extremely intelligent, yet simple and sincere individual, who toward the end of his life may have been under more stress than he could cope with.

A Generator Instead of a Compass

Lester Hendershot didn't set out to invent a new energy device. In the early 1920s, he was interested in making a better compass when he built a small device that interacted with the earth's magnetic field. To his surprise, his device spun like a motor. Hendershot decided that the earth's turning created friction against its magnetic field, much as a spinning ball covered by a cloth would create friction with the cloth, and that his device was able to capture the electricity generated by this friction. As he tinkered with his discovery, he eventually ended up with a device that put out enough power to run a tabletop radio and a 120-volt lightbulb at the same time.

Hendershot first took his invention to the manager of Bettis Field, a nearby airstrip. That's where aviator Charles Lindbergh first saw the device, and became interested in its development. Eventually, Hendershot and his generator went to Selfridge Field in Detroit for its first significant experiments.

News of Hendershot's device leaped onto the front pages of the nation's newspapers in late February 1928 when Lindbergh witnessed a demonstration at Selfridge, along with the airfield's commandant, Major Thomas Lanphier. Newspapers of the time said that powerful groups of financiers were extremely interested in the invention. Other news articles said technicians at the field had built a Hendershot device under Lanphier's orders and Hendershot's supervision. Because of the Lindbergh connection, Hendershot was quoted in newspapers for days that year.

Hendershot Is Silenced

And then, suddenly, for Hendershot and his device—oblivion. As the newspapers left the matter in March 1928, Hendershot had received a powerful electric shock while demonstrating his invention at the Patent Office, and was hospitalized. The press said it was a jolting 2,000 volts, but Hendershot told his family that he had been zapped by only 220 volts, according to his son, Mark. Since Lester's vocal chords were temporarily paralyzed, he had to recuperate for several weeks.

Mark says that while his father lay in the hospital bed, a corporate representative managed to get Lester to agree that he would not work on his invention for twenty years, and that he was paid $25,000 on that condition. Lester never revealed which large corporation had paid him.

Lester Hendershot then dropped out of public view for decades, until Ed Skilling found him in the 1950s through a mutual acquaintance. Skilling and the acquaintance took Hendershot's unusual device to Skilling's laboratory, but failed to get it working. Skilling returned the unit, figuring that he would just dump the project and file it away as a hoax.

It didn't turn out that way. Before Skilling left the Hendershot home, seven-year-old Mark fiddled with a tuning knob until the lightbulb they were using to indicate output flashed on. Skilling had searched for hidden batteries and knew there were none, so he knew that the bulb was lit only by energy coming through the machine itself. Thus, Skilling stayed on the project. However, only the Hendershots ever made the device work, and Lester did not know how to duplicate it or scale it up. "I'm not one of these slide-rule boys," he said. "It's pretty much cut-and-try."

Mark Hendershot recalls being told that his father received a phone call on April 19, 1961, from a man with impressive credentials who said he would be able to get financing. But before that could happen, on that same day, Mark returned home after school and found his father dead in the family car, the engine running and the exhaust piped through a window. It was recorded as a suicide without any further investigation. Mark finds this hard to believe, especially since the phone call preceded Lester's death by only an hour or so. But Mark has never been able to come up with any evidence of foul play.

Skilling regrets that Hendershot did not live to meet T. Henry Moray, "as the combination of the Hendershot simplicity of circuitry with Moray's knowledge and theory of Radiant Energy would astound mankind." Mark Hendershot, now a Vietnam veteran and family man with an electrical contracting business in Washington state, just wants to see the record straight. Newspapers of the past had left the impression that the Hendershot device was a hoax. But Mark—who continues his father's work—asks if that is so, then why did a man of Lindbergh's stature give the hospitalized Lester Hendershot an expensive silk smoking jacket as a get-well gift?

VIKTOR SCHAUBERGER AND THE SPIRAL OF ENERGY

Another inventor who came to a sad end is Austrian Viktor Schauberger.

As a forester early in this century, Schauberger (1885–1958) had spent countless hours watching vortexian turbulence—three-dimensional spiralling—of the water in wild rivers. His employer at the time was an Austrian prince, and the royal family owned a huge area of untouched wilderness. This gave Schauberger years to study the life processes in the mountains under his care.

The Energized Vortex

According to one of his biographers, Schauberger saw unusual sights in this pristine ecosystem, such as a landlocked lake renewing itself with a whirlpool, followed by a giant waterspout. At nights, by a waterfall in the light of a full moon, he learned about the heightened energy state of cold water by seeing rocks float.

The theme he saw in nature's movements and designs was the vortex, a type of spiral (see "Spirals of Energy" on page 13). Keeping in mind his motto of "understand nature, then copy nature," the observant genius made what he called "living machines." Today's main energy technologies use outward-moving *explosion*, such as fuel-burning and atom-splitting. By contrast, Schauberger's machines operated on principles of inward-spiralling movements of *implosion*. In short, he had discovered how to generate electric energy in a radically different way by working in harmony with nature's creative movements. We'll discuss the principles behind Schauberger's device more thoroughly in Chapter 11, when we'll meet a man who has continued his work.

Why Was Schauberger Suppressed?

Recent discoveries by Schauberger's biographers and by new-energy researchers shed some light on what happened to the inventor, although there are still unanswered questions. In 1958, when he was seventy-three years old, two Americans persuaded Viktor and his son Walter to go to the United States. The Nazis had forced Viktor to work on his energy-generating device in a prison camp—or else say goodbye forever to his family. Now, a consortium was promising to manufacture his beneficial energy devices. It was something that he had always wanted.

That visit to America turned out to be an ordeal in a sweltering Texas summer. An atomic energy expert came down from New York, met for three days with the Schaubergers, and reportedly wrote in a document viewed by them that Viktor was correct—that his biotechnology was the path of the future. But the Schaubergers' hosts soon revealed their insincerity—they were not in any hurry to develop his generator.

In order to be returned home, Viktor had signed a contract during his stay in the United States that forbade him to ever write about or even talk about his past or future discoveries. The consortium now owned all the rights to his implosion-generator secrets. When father and son stepped on an airplane to go back to Austria that fall, Viktor was broken in spirit and Walter was filled with a bitterness toward the United States that lasted throughout his life.

On the way home, Viktor cried repeatedly, "They took everything from me, everything. I don't even own myself." Five days after they returned home, he died, heartbroken. Instead of being rewarded for his work, Viktor Schauberger's life ended in despair.

Today's researchers find a cold trail when they try to discover what happened to Viktor Schauberger's papers detailing the non-polluting generator. For example, Erwin Krieger of Ohio, a retired industrial scientist, ran into a wall in 1993 when he petitioned, under the Freedom of Information Act, for papers connected with the "interrogation or debriefing" of Viktor and Walter Schauberger in Texas, July to September 1958. Krieger's request for information was denied by the Central Intelligence Agency. The CIA neither confirmed nor denied the existence of the papers, citing the National Security Act of 1947.

Why would anyone trick Viktor Schauberger into thinking that they would bring his knowledge out to the world, only to bury the knowledge once they had it? One speculation comes from a nuclear physicist and electrical engineer who has participated in the alternative energy field longer than nearly anyone else. Dan A. Davidson of Arizona writes about new-energy devices in general:

> Various power groups know that if mankind has unlimited energy at its disposal, it becomes virtually impossible to control and manipulate people. With free energy a person is not subject to those who would control his transportation via fuel

curtailment. One could live virtually anywhere since a readily available energy supply could be used to make any environment liveable. Water could be taken from the air via condensation if necessary; and with water, food could be grown. With *unlimited* energy available to a country, it could synthesize anything, including the atomic elements; therefore, that country is not open to international blackmail because of energy resource requirements. Stated in brief: ENERGY = FREEDOM. [Emphasis in original]

WILHELM REICH AND THE ORGONE MOTOR

Like Walter Russell, the inventor of another new-energy device was a highly educated man. Wilhelm Reich (1897–1957) was an Austrian—later an American—scientist and outspoken innovator in fields ranging from psychiatry to biology. All of his work led toward one unifying discovery—a pulsating, life-force energy found everywhere, in varying degrees. Reich named this energy "orgone," because he first discovered it in living organisms.

In 1948, the famed educator A. S. Neill of England saw a small motor running in a workshop in Maine. It was connected only to an "orgone accumulator," without any other source of power. His friend, Wilhelm Reich, was the proud inventor of the setup. "The power of the future," Reich said.

Why didn't Reich, a prolific discoverer, continue this line of research? "My job is discovery, and I leave it to to others to carry out the results," he said in a letter to Neill.

No one took up Reich's work, and he died in a federal prison, his books and papers burned. Reich's life ended after prolonged conflict with the Food and Drug Administration, during which the FDA gathered evidence for a case against the use of his orgone accumulator in physical therapy.

Static Electricity: Overlooked Free Energy

Although Reich first discovered orgone while doing research in psychiatry and biology, he found that it could be used as a motor force. In 1947, he bought a Geiger counter to detect cosmic radiation, since he thought orgone might have properties similar to the cosmic rays that constantly stream into our atmosphere from space. When he put the counter inside an orgone accumulator—a box designed to capture and concentrate orgone—the counter registered the normal background radiation by clicking at a normal rate, about thirty counts per minute. Reich then turned to other projects, and stored the Geiger counter next to a miniature orgone accumulator.

A few months later, he picked up the Geiger counter and tried it again, and found it was clicking along at an amazing 6,000 counts per minute. After performing some tests, Reich decided the Geiger counter had become saturated with orgone energy. A year later, he found that vacuum tubes—the kinds of tubes that used to be found in television sets—were also affected by long soaking in the concentrated orgone environment. These tubes showed the strong effects of orgone by giving off a strong violet-blue light. This led Reich to try using the tubes to run a motor, a feat that was witnessed by five members of his research staff.

Based on these and other experiments, Reich decided that static electricity and orgone are related. Static electricity is electricity at rest, as opposed to dynamic electricity, which flows in a current— the electricity that causes hair to stick to your comb in the winter versus the electricity that flows through the wires in your home. Reich thought that orgone was the one primary energy, similar to static electricity in that it permeates large areas without irritating the living beings in those areas. Dynamic electricity, by contrast, is a coarser form of energy that irritates living beings.

Such a conclusion—that orgone and static electricity are related— could answer some of the unanswered questions in new-energy science. For example, electrostatic motors—motors that run on the energy in the space surrounding them—are being rediscovered. These motors fell into disuse after the discovery of capacitors, devices that could store electric charge taken from the modern-day power grid. Dan Davidson and others see static electricity as being a major key to new-energy systems in that the source of power is freely available—you don't have to plug into a wall socket.

Reich's Books Are Burned

One reason that Reich did not pursue the development of the orgone motor was his ongoing fight with the Food and Drug Administration, which took up his time and energy for years. In 1954, the FDA, in an attempt to suppress the use of the orgone accumulator in physical therapy, ordered Reich's hardcover books banned from circulation. His softcover books, including all his periodicals, were burned by government employees. Anything that mentioned the word "orgone" literally went to blazes. For refusing to obey an injunction against publishing his material, Reich was sentenced to two years in jail. He died in prison in 1957.

In his long and prolific career, Reich voiced many unusual observations, especially when under stress toward the end of his life, such as his claim that his "cloud-buster" machine could affect UFOs. Therefore, he has been another easy target for skeptics who focus on an innovator's amusing eccentricities rather than on his accomplishments. Is there another way to look at such innovators? One satirical but tolerant author, Donna Kossy, observes that "ideas discredited by powerful institutions are often driven underground to the realm of kookdom."

A few new-energy researchers have tried to build the orgone motor, but found that Reich left inadequate notes. One inventor was so intent on getting information that he broke into the Wilhelm Reich Museum in Rangeley, Maine. He was imprisoned for burglary, and the information returned to the museum. Other researchers are looking into orgone's medical effects.

The next part looks at some of today's energy innovators, starting with a chapter that gives numerous views of what seems to be one energy source—the background sea of energy in the universe.

Part II
Space Energy
and the New Physics

The imaginations of the people who design and build our world, from electric power plants to automobiles, are generally limited by the prevailing body of knowledge of how things work. Most engineers, if not taught about a source of energy, will not invent devices that use it. In this section, we will meet some of the inventors who have leaped beyond the limits of today's knowledge.

First, though, we will look at their emerging worldview, a worldview that I believe will spread from physics to the other sciences. It is, in a sense, an old worldview. A century ago, science was outgrowing the old idea of a motionless source of energy called the aether, which was thought to fill space like a liquid that had long ago been poured into the universe. An experiment that failed to detect such an aether convinced many scientists to throw out their belief in an energy that permeated all space. Albert Einstein pulled together a complex theory, his theory of relativity, that explained how an aetherless universe could work.

But now, the new-energy researchers have returned to an aether-based theory of the universe. But their theory has a twist—it is based on an aether in motion. We will see why these researchers have returned to this idea, and then see how this new theory has been used to create various new-energy devices.

4
A New Physics
for a New Energy Source

*Today the vacuum [of space] is not regarded as empty. . . . It is a
sea of dynamic energy . . . like the spray of foam near a turbulent
waterfall.*

—Harold Puthoff,
Physicist

*Now we are confident that the universe is formed from non-materi-
al primary substance, which may be described as the shadow charge
that gives birth to all things.*

—Shiuji Inomata and Yoshiyuki Mita,
Researchers

Moray B. King, a graduate student at the University of
Pennsylvania, risked upsetting a committee of engineering
professors in 1978 with his proposal for a doctoral thesis topic—
that energy could be tapped from space. The personable, good-
humored King had not set out to shock. In fact, as a dutiful systems
engineering student, he had at first accepted the standard view that
the vacuum of space is useless as an energy source.

However, King had become intrigued with a new idea a couple of
summers before, after reading a book about UFOs. In his search
through the physics literature for principles that would allow for
antigravity, he ran across a concept that interested him even more—
something called "zero-point energy." It not only allowed for antigravity, it also allowed for an abundant source of energy.

WHAT DO THE TEXTBOOKS SAY?

Most scientists and engineers have been taught that the vacuum of space is completely empty and still in the absence of heat, light, and matter. Unless a student is studying quantum mechanics, his or her textbook never mentions zero-point energy.

The quantum mechanics student does learn that the fabric of space consists of random fluctuations of electricity. He or she also learns that these fluctuations are collectively called zero-point energy because they represent the energy that is present even at a temperature of absolute zero, the temperature at which everything is completely cold. It is the energy that exists when all other sources of energy are taken away.

This energy is difficult to detect because it is everywhere. Expecting someone to sense it is like asking a fish to detect the ocean; the fish has no concept of a world that isn't an ocean. Similarly, the fluctuations of electricity that make up space energy are too microscopic and too quick for us to sense them, either with our bodies or with standard detection equipment.

Why did Moray King's engineering professors fail to teach him about zero-point energy—what we refer to in this book as space energy? The reason is that scientists assume that these vacuum fluctuations simply even out. They call this the second law of thermodynamics, also known as the law of entropy. Under this law, everything is doomed to increasing disorder, until all comes to a dead rest. This means that, according to traditional science, space energy cannot be put to any practical purpose because its randomness cannot be made into an organized system. It would be as if a pile of threads suddenly organized themselves into a shirt.

A NEW ENERGY PHYSICS:
MAKING THE IMPOSSIBLE POSSIBLE?

King had found the most impressive reference to zero-point or space energy in a book called *Geometrodynamics*. The author, noted physicist John Archibald Wheeler, said that this energy foaming within the fabric of space is enormously powerful—if formed into an object, it would put out more energy than a bright star. That's a lot of power.

Does this source of incredible power really interact with our world? King found that there, too, the physics literature held good news.

Quantum mechanics—the branch of science that deals with protons, electrons, and other basic particles of matter—teaches that super-high-frequency energy does interact with physical matter all the time. It says that these basic particles are mixed with space energy.

The difference between standard quantum mechanics and the ideas of Wheeler and other scientists is that they believed basic particles such as protons and electrons were not only mixed with space energy, they were actually *made out of* space energy. As King continued to read through books on the subject, he began to see energy as a flow, a river from another dimension of space, and basic particles as tiny whirlpools in that river. If the river ceased to flow, the basic particles—the building blocks of all matter—would disappear. So would everyone and everything.

Filled with a sense of awe, King began to see beyond the standard view of space energy as a random jittering of basic particles. He found his newfound ideas confirmed by the work of Timothy Boyer, Ph.D., a physicist and teacher. Boyer said—contrary to traditional scientific belief—that space energy did influence matter, the physical world around us, and that it wasn't random and meaningless.

Eventually, King realized that if engineers could get only a small part of those random energetic movements in space to line up with each other, they could tap into a tremendous source of power for our everyday world.

COMING UP WITH A NEW COMBINATION OF THEORIES

King wondered: Why was no one asking if all that power could be harnessed and put to work? The answer seemed to lie in specialization. The people who make machinery and generators to move and heat and power things—the engineers—don't necessarily study quantum mechanics. The people who do study quantum mechanics, the ones who come up with the equations and the formulas—the physicists—don't build machinery.

Even if the majority of neither the engineers nor the physicists were interested in this topic, King was. He still wanted to find out if there was a way to allow for the use of space energy. So the young student set a task for himself. He would stick to the standard physics literature and look for concepts that could be put together to form a body of knowledge—a combined theory—about the feasibility of tapping into that abundant energy. He searched through the respected journals and found articles that, taken together, made a case for doing what his professors had said was impossible.

Academia was not particularly interested in space energy at the time, but a growing audience, mostly outside of the ivy walls, snapped up the book that King eventually wrote. *Tapping the Zero-Point Energy*, first published in 1989, brought together published theories about space energy and theories about the way that natural systems organize themselves. This book laid the groundwork for the development of a coherent theory behind a new source of energy.

From Chaos To Order

Russian-born scientist Ilya Prigogine won a Nobel Prize in 1977 for showing how certain systems can evolve from random behavior to orderly behavior. This means that entropy, which assumes that all systems become more and more disordered, is no longer the only game in the universe. It means that energy can indeed be seen as a creative force in space, instead of disorganized chaos. This behavior, the opposite of entropy, has since been called negentropy.

Dr. Moray B. King
of Utah wrote one of
the first books that
explained the existence
of space energy.

In the 1970s, before and after graduation, Moray King started keeping a foot in two worlds, one in the world of theoretical physics and one in the world of the practical tinkerers who were trying to capture space energy in their home workshops. He was introduced to that second world by new-energy author Christopher Bird, who told King about T. Henry Moray and Moray's struggles to capture space energy (see Chapter 3).

New-energy ideas hit King from all sides after that. At first, he wondered if he was being introduced to a bunch of kooks, but he soon came to appreciate these concepts. He kept asking questions, networking and presenting papers at conferences on new-energy technologies, and encouraging inventors to come up with a reproducible experiment to prove that space energy could be tapped.

By 1994, King had further refined his space-energy ideas. At conferences, he explained to eager audiences how vortices—whirlpool- or tornado-like spirals found everywhere in nature—held a key to the energy lock. Give a sudden twist to the nucleus of an atom, and all its neighbors, and keep spinning them, King said, and you may pull some space energy into your electric-generating system. Rotate the spinning materials—a spin upon a spin—and you have a better chance of picking up some extra energy. Then build pairs of counter-rotating vortices into your system, and you would really have something.

To partially visualize this concept, you can take two yo-yos, twist their strings, and let go so that both yo-yos start spinning. You can then swing the yo-yos in full circles, one forward and one backward. This is the type of motion that might give an inventor a chance to hit the space-energy jackpot.

An Ancient Idea is Reexamined

To more fully understand King's ideas, it is helpful to go back to a very old concept. Another way of speaking about the background sea of energy is the ancient term *prana*, later known as the *aether*. In the eighteenth and nineteenth centuries, the aether was considered to be a substance filling all space, and through which light travelled.

In 1887, two Americans—Albert Michelson and Edward Williams Morley—tried to detect the aether experimentally. They could not, and concluded that the aether did not exist. About thirty years later, the concept was totally discarded when Albert Einstein put forth his theory of relativity. It says that there is no background structure to the universe, such as an aether. Instead, all objects in the universe, such as stars and planets, affect each other. This means that nothing in space is absolute.

But, as with all theories, there were things that Einstein's theory could not account for. So in 1954, distinguished English physicist P.A.M. Dirac asked science to take another look for the aether: "The aetherless basis of physical theory may have reached the end of its capabilities, and we see in the aether a new hope for the future."

An American scientist, E. W. Silvertooth of Washington state, responded to Dirac's call. In 1986, Silvertooth performed an experiment using laser equipment and his knowledge of advanced optics. By measuring earth's motion in space, he calculated that our solar system is moving toward the constellation of Leo at nearly 400 kilometers a second—or about 892,800 miles an hour. Silvertooth had succeeded where Michelson and Morley had failed. The fact that earth's motion in space could be detected meant that there had to be a stable point of reference—such as the aether—for this motion to be measured against.

In order for a scientific experiment to be considered valid, it must be successfully repeated. However, Silvertooth used some very expensive equipment, and his research was sponsored in part by the United States Air Force and another defense agency that handles advanced research. To my knowledge, Silvertooth's experiment has not been repeated, although an Austrian physicist has claimed to have also detected the aether.

A Fast-Spinning Vortex?

Today's aether theorists do not see the aether as an invisible fluid filling all of space. Rather, they say it is a spiralling foundation for the universe that cannot be detected by current measuring instruments because its movements are too quick.

Moray King is not the only space-energy scientist who thinks that the aether moves in a spiral motion. Paramahamsa Tewari, Ph.D., of India is another. He says that the idea of there being enormous levels of power for every square inch of space would be wrong unless that space rotates at a fantastic speed, "like a vortex." He sees the universe as being in motion according to its basic design, with just a concentration of matter here and there—a galaxy, a solar system, a planet, an electron.

What makes this movement difficult to detect is the fact that we are moving along with it, and thus have nothing to compare it with. It is like trying to sense the spinning of the earth on its axis—because everything is spinning, including us, we do not feel the motion. One scientist describes space energy as two giant, invisible elephants pushing on both sides of a door. As long as they push with equal force, the door does not move one way or the other.

The aether not only exists, the space energy it produces energizes the earth. To understand how it works, think of a microwave oven. If you put a potato in a microwave, you do not see it cooking, nor do you feel any heat coming from the oven. That's because the microwave cooks the food from the inside out. The oven remains cool, but the inside of the potato becomes very hot. In the same manner, space energy "cooks" the earth's core, which is very hot, while the earth's surface remains relatively cool. The big difference is that the energy in a microwave comes from outward-moving forces of decay, explosion, or combustion, while space energy takes the form of an inward-moving spiral—as explained in "Spirals of Energy" on page 13.

Despite a theory that supports a universal abundance of space energy, many engineers cannot let go of their belief in a world ruled by a finite amount of energy. To be fair to the engineers, they don't want to give up that belief because it has worked well as a basis for practical engineering. It is the idea at the heart of the Industrial Age.

However, the new-energy theorists say that space energy does not violate the laws of conservation of energy, which state that energy can be neither created nor destroyed. According to these theorists, this energy has always existed, and thus is not being created out of nothing. It can simply be put to human use. "People are having trouble deciding whether they want to believe it or not," King says.

MAGNETS AND ENERGY

The key to many of the devices that you will be reading about is the magnet. The earth's own magnetic field—the one that makes a compass point north—may interact somehow with space energy. And new-energy researchers find that the smaller magnetic fields which surround manufactured magnets play a key role in getting their energy-generating hardware to work. Some inventors use super-powerful magnets made of rare materials, while others use the sorts of ordinary magnets that are found in stereo sound systems.

How exactly do magnets tap into space energy? It is not possible to answer that question with any authority, since scientists are unable to explain exactly what a magnet's force field is—the force that attracts metal objects to the magnet. Nor can they explain exactly what that field interacts with. One electronics engineer says that we are like early humans discovering fire; they knew what it did, but they didn't know why. Many new-energy researchers have come up with differing theories of what makes magnets work. But these theories have not yet jelled into one body of knowledge that is accepted by the scientific establishment.

One thing we do know about magnetism is that it is related to electricity. In the 1830s, English scientist Michael Faraday showed that magnets could be used to produce electricity, and that an electric current produces a magnetic field. While it is not fully understood why this happens, this knowledge has been put to practical use in electric motors and generators. So it is not surprising—if in fact space energy is electric in nature—that magnets can be used to capture space energy, even through we don't fully understand how.

MAVERICKS IN HIGH PLACES

In the past decade, Moray King has been joined by scientists around the world in space-energy research, and their results have caused great excitement in the new-energy world. Former astronaut Edgar D. Mitchell, Ph.D., predicted this excitement in 1980 when he said:

> There are types of energy which lie outside the electromagnetic spectrum. Unfortunately, these research efforts have not been given recognition. For the most part, they have been performed by individuals . . . without any support, whose work lies at the threshold of present-day science, and who are years ahead of science which is already established.

The fact that many of space energy's newer proponents are people who have been part of the science establishment means that space energy, long thought of as an oddball idea, will have to be taken seriously.

Dr. Harold Puthoff
of the Institute for Advanced
Studies in Austin, Texas,
is an important
space-energy theorist.

Harold Puthoff, Ph.D., of the Institute for Advanced Studies in Austin, Texas, is giving space energy the publicity that Mitchell thought it lacked. Puthoff is a scientist whose low-key personality fits into a variety of settings, from security-clearance laboratories to meetings of environmentalists. His background includes corporate work, several years with the United States Department of Defense, and a stint with Stanford Research Institute International. He gives briefings to top government officials and oil industry executives, and to other audiences worldwide.

Puthoff was named Theorist of the Year in 1994 by *New Energy News*, for a paper that *News* editor Hal Fox, Ph.D., called the century's most important theoretical paper. Puthoff and two coauthors say that inertia—the tendency of a body in motion to remain in motion, or a body at rest to remain at rest—can be explained by the presence of space energy. Puthoff explains by saying it is space energy that knocks you down if you are standing on a train and the train accelerates quickly from a full stop.

Fox says, "The way the various institutions of science are structured, it is important to work within the system to successfully introduce new scientific theories and facts. This is what Dr. Harold Puthoff has gently accomplished over the past few years."

Thomas Bearden, a retired United States Army lieutenant colonel, is a more controversial theorist who is considered to be almost

Ret. Lt. Col. Thomas
Bearden of Huntsville,
Alabama, has worked
out a way to explain
how space energy
can work under the
laws of physics.

a guru by some in the space-energy field. Bearden believes that present-day mechanical and electrical engineering concepts and mathematics are based on the manipulation of effects, and not of underlying causes, in the same way that a driver can accelerate and deaccelerate a car without understanding how an engine works. The devices made by mainstream engineers do the work they are intended for, he notes, but are crude compared to the hardware that could be made if the deeper causes were understood.

Bearden's quest parallels King's—to learn how to create order in a small part of the seething vacuum of space and put that tremendous energy to work: "We can dip our paddlewheel into that river."

Puthoff and Bearden are only two of the many conventionally trained scientists who have found in space-energy theory a new way of seeing the world. And their ideas of theoretical physics are not only important to the world of science. Their ideas form the basis for a technology that will ultimately affect everyone.

In the next chapter we meet inventors who have tried to turn space-energy theory into space-energy devices.

5

Solid-State Energy Devices and Their Inventors

Imagine a world in which endless, nonpolluting, and virtually free energy powers our cities, cars, and homes.

—Owen Davies,
Science writer

Our electrical company tells us that the only two practical choices for their power are coal or nuclear. There is another alternative.

—Wingate Lambertson,
Inventor

In this chapter, we'll meet three of the leading North American inventors of solid-state energy devices, or devices that use no moving parts. These inventors are only three of many.

These men have diverse backgrounds and personalities. In California, a scientist described by *Omni* magazine as a star in the electronics field works in a high-tech private laboratory funded by financial backers. In Florida, a former government official pays for his research out of his retirement savings, and makes discoveries in his garage. In Canada, a self-described eccentric, well-known in Japan but unknown in his own country, cooks up a crystal-based energy device in a tiny kitchen—using ordinary rocks.

What these inventors have in common is a zest for exploring. Their work on the leading edge of energy science holds promise for the development of small-scale, quiet but powerful converters—devices that convert space energy into useable electric power.

THE CHARGE CLUSTERS OF KEN SHOULDERS

Ken Shoulders, Ph.D., a tall, solidly built man, wears the expression of someone not inclined toward ordinary concerns. He is a discoverer on the frontier, and lets others worry about whether his findings fit into the accepted boundaries of scientific theory.

In the early 1960s, Shoulders developed much of today's microcircuit technology. Now, he is working on an even more advanced concept: the high-density charge cluster. It is a concept that holds great promise in the space-energy field, since these donut-shaped, microscopic clusters put out more than thirty times the energy required to produce them.

Shoulders spent decades doing work in various institutions, wherever he had a chance to learn more about science and to try things out. This work included nonteaching staff positions at universities such as Massachusetts Institute of Technology, in laboratories such as Stanford Research Institute, and in private corporations. Along the way, Shoulders accumulated the equipment he needed to set up his own laboratory, which he did in 1968.

Like Nikola Tesla, the father of new energy we met in Chapter 2, Shoulders made a discovery that could render his previous work in microcircuit technology obsolete. It was a discovery made by accident.

Around 1980, Shoulders was introduced by physicists at the Stevens Institute in Hoboken, New Jersey, to strange strings of particles—what scientists call vortex filaments. After working on them for awhile, Shoulders found that they weren't strings at all, being about as broad as they were long. They showed up as strings on the instruments of most researchers because the researchers could never stop the motion of these extremely fast-moving blobs. When Shoulders learned how to get clear pictures of the blobs, he found they were little beadlike structures. The simplest name for them is charge cluster, although Shoulders calls them Electrum Validum, a name that means "strong charge."

What Is a Charge Cluster?

The basic idea of a charge cluster is rather simple. It is a tightly packed cluster of about 100 million electrons, an electron being the part of an atom that revolves around the nucleus. Shoulders has been able to create conditions under which electrons break free from their nuclei and join together into remarkably stable little ring-shaped clusters, like tiny donuts. "It is the wildest electronic effect you will ever see," Shoulders says, calling his creations "little engines of vast complexity that just don't die!"

As simple as the charge cluster is, conventional science has a hard time accepting its existence. That's because it violates a law of physics: "Like electrical charges, either negative or positive, repel." Since all electrons carry a negative charge, conventional science says that they should not cluster.

Hal Puthoff, whom we met in Chapter 4, has worked with charge clusters, and thinks that the force which holds them together is the result of an effect named after Dutch physicist Hendrik Casimir. The Casimir effect refers to the tendency for two perfectly smooth metal surfaces placed near each other to come closer together. Puthoff explains the effect this way: imagine two metal plates hovering in space, close to each other. Because the plates shield each other from space energy coming from one direction, the space energy pressing in on each plate from the opposite direction would slam the two of them together, releasing energy as heat.

Shoulders uses the Casimir effect to pinch a cold plasma—a special form of gas that conducts electricity—to create heat and charge clusters. The electricity he uses is static electricity, the electricity in the spark that snaps from a doorknob if you drag your feet across a carpet. In Shoulders's system, this electricity provides the electrons that make up the cluster. It is, essentially, an electric charge compressed into a visible form.

What inspires Shoulders's awe about these tiny entities is that they almost seem to have an intelligence about them—they are self-organizing. The clusters appear to form into various sizes, but are uniform in organization and behavior. They often look like a ring or a necklace of tiny donuts. "It's some law of nature that's just not spelled out for us yet," Shoulders says.

Shoulders discovered the link between charge clusters and space energy when he tried to find out what could supply the large amounts of energy needed to make electrons overcome their tendency to repel one another and join into tightly packed clusters. Their high energy makes charge clusters very powerful—they can bore holes through ceramic tile without losing strength. Because of the Casimir effect, space energy appears to fit the evidence from Shoulders's experiments as a likely source of this energy.

As futuristic as this technology seems, Shoulders has been able to convince a tough customer of its value—the United States Patent Office. While past attempts to base a patent on space energy have been unsuccessful, Shoulders has broken through with a 1991 patent titled, "Energy Conversion Using High Charge Density." It is a milestone—the first successful patent to say that space energy can be used as a source of practical electric energy.

Charge Clusters and Commercial Products

Now working with his son, Steve, Ken Shoulders continues to make breakthroughs. What Shoulders sees under the microscope is another world, hinting of future machines that will be thousands of times more powerful than our current machines.

Charge-cluster technology could be one of the first space-energy technologies to be commercialized. Unlike some of the other space-energy inventions, charge clusters do not need magnetic fields or low temperatures to work. One new-energy writer says the charge cluster may be one of the most promising areas of research since the transistor.

Providing abundant clean energy is not the only thing that charge clusters can do. There is a whole range of possible products based on charge-cluster technology, according to Puthoff, who lists a few of the products besides energy devices that could result from developments in this field:

- High-resolution television screens flat enough to hang on a wall.
- Notebook computers more powerful than the largest mainframe.
- Tiny X-ray machines that can enter the body and kill cancer cells without harming surrounding tissues.

While the Shoulders team makes advances in the laboratory, a private firm with the necessary product-placement know-how makes plans in the marketplace. This firm will ensure that charge-cluster technology can be licensed worldwide for eventual development into a number of products.

THE CERMET OF WINGATE LAMBERTSON

In Florida, Wingate Lambertson, Ph.D., lights a row of lamps in his garage using what he says is electricity taken from the energy of space. It took years for Lambertson, a former director of Kentucky's

Science and Technology Commission, to overcome his academic skepticism about claims that you could get something for nothing—that energy freely available from space could be tapped for useful work.

After getting his doctorate from Rutgers University, Lambertson worked for United States Steel in Chicago before going into the United States Navy. After going back to Rutgers for more postgraduate work, he joined Argonne National Laboratory, where he worked on nuclear fuel technology.

Then Lambertson discovered the large body of space-energy literature that has been written by researchers in the field. Eventually, he came to believe that something similar to an aether—the basic stuff of the universe discussed in Chapter 4—could exist, and that when collected, it could be used to make electricity.

After more than two decades of research and experimentation, Lambertson is certain that space energy can be turned into a practical power source through a process he calls World Into Neutrinos (WIN). He envisions it being engineered into units that will probably be set outside the home on a small concrete pad, like central air conditioning units are now, and wired into the home's master electric switchbox. The price? About $3,000 for either sale or lease—cheaper than buying or leasing a car.

Dr. Wingate Lambertson of Florida holds the energy-gathering part of his E-dam, which uses moving electrons to capture space energy. This part is made of cermet, a material made of ceramic and metal mixed together.

The WIN Process and Cermet

The most important part of the WIN process is Lambertson's E-dam, and the most interesting component in the E-dam is cermet. Cermet is a heat-resistant ceramic-and-metal composite invented in 1948 and considered by NASA for rocket nozzles and jet-engine turbine blades. Lambertson, who spent almost his entire career working with advanced ceramics, is experimenting to develop the best cermet for his device. The E-dam contains a plate of cermet formed into a round spacer about three inches in diameter, sandwiched between metal plates of the same size.

The process starts with an electrical charge—basically, a stream of electrons—from a standard power supply. The charge flows into the E-dam, where it is held in the cermet: "It stores electrons like a [regular] dam stores water," Lambertson says. When the dam is opened, the electrons are released. As they accelerate, the falling electrons gain energy from the space energy that is present in the E-dam. This gain in energy is what allows the device to put out more power than it takes in.

The current of electrons then flows into the device to be powered, such as a lamp, and then moves into another E-dam for recycling. Lambertson says there is no way for the process to become dangerous—if too much power were generated, the E-dams would overheat, shutting down the system.

For years, Lambertson was more interested in proving that the process gained energy than in the actual amount of energy gained, since he thought scaling up the process to higher efficiencies would be a relatively simple engineering problem. When his first of three patent applications was rejected, he saw it as a blessing because it forced him to study the space-energy literature more carefully. By the fall of 1994, he had improved the process to the point where it put out twice as much energy as it started with.

Lambertson Finds Help

Meanwhile, Lamberston was having a frustrating time in trying to find funding and marketing help. Responses to his proposals usually fell into one of two categories:

• "This will not work, your calculations are in error."

• "You get it working and free of all technical problems, and we will take it off your hands."

He learned, as have other inventors in this book, that it's a waste of time to try to convince people of the validity of one's claims when those people don't want to listen. But he did find support in 1987, when he spoke at a new-energy conference in Germany. There, he found people who saw the need for his invention and agreed to market it when the WIN process is perfected.

Lambertson says that he now has active associates in Switzerland, in addition to interest shown by the United States Navy. Three different groups have shown interest in taking over and developing the WIN method.

THE DIRT CHEAP ROCKS OF JOHN HUTCHISON

If you ask the other residents of a certain apartment building in Vancouver, they may admit to being curious about John Hutchison. They see a tall, muscular man who carts old consoles of electronic equipment onto the elevator nearly every week. Their curiosity increased the day a Japanese television crew showed up and disappeared inside his apartment for a few hours. And in the summer of 1995, Hutchison further puzzled onlookers by sitting on the curb

John Hutchison
of Vancouver has
created devices that
use crystals—including
crystals made from
ordinary rocks—
to capture space
energy.

and picking out stones. Why would a rockhound sort through ordinary street rocks?

What the neighbors do not know is that John Hutchison is well-known in new-energy circles, and is even known to some who move in the circles of established science. His visitors have included distinguished physicists. But unlike Shoulders and Lambertson, he is a self-taught scientist. As a boy in Vancouver, he read about Nikola Tesla (see Chapter 2) and then startled neighbors with Tesla coil experiments in his backyard.

While in his twenties, he developed a medical problem that resulted in his living on a small disability pension. For years, he lived a generally reclusive life, digging for rare electrical equipment in military surplus stores and junkyards, and carrying his finds home on the city bus. Apart from time spent as a volunteer at a local ecology center, he spent hours in his bedroom-turned-laboratory, patiently rebuilding equipment. He considered opening a museum.

Antigravity and the Hutchison Effect

Hutchison's life changed drastically in 1979 when, upon starting up an array of high-voltage equipment, he felt something hit his shoulder. He threw the piece of metal back to where it seemed to have originated, and it flew up and hit him again. This was how he originally discovered the Hutchison effect. When his Tesla coils, electrostatic generator, and other equipment created a complex electromagnetic field, heavy pieces of metal levitated and shot toward the ceiling, and some pieces shredded.

What is the Hutchison effect? As with much of the new-energy field, no one can say for sure. Some theorists think the effect is the result of opposing electromagnetic fields cancelling each other out, creating a powerful flow of space energy.

A Vancouver businessman heard about the Hutchison effect, contacted Hutchison, and brought in a consulting engineer to form a company that would promote technology developed from the effect. Despite demonstrations to potential customers from both Canada and the United States, things did not work out, and Hutchison and the company parted ways in 1986.

After a couple of other abortive business tries, including a sojourn in Germany, Hutchison returned to Vancouver in late 1990 and again lived a relatively reclusive life. Piece by piece, he sold what remained of his laboratory equipment in order to pay his bills. It would be several years before he could reestablish his collection.

Hutchison wanted to connect with other researchers, but the local media had given his work the weird-science treatment, and didn't take him seriously. However, material on the Hutchison effect was included in a Japanese book on Hutchison's life and work that sold well in Japan. Living in a country with almost no natural resources has led the Japanese to take new-energy ideas very seriously, as we will see in Chapter 8.

As a result, Hutchison was asked to speak in Japan, where thousands of people paid to attend his two lecture tours. These tours were organized by Hiroshi Yamabe, a well-known Tesla lecturer who made his fortune in such advanced engineering fields as robotics and artificial intelligence. Yamabe offered to set up a laboratory for Hutchison, but the Canadian was ambivalent about the prospect of moving to Japan.

Beyond the Hutchison Effect:
The Dirt Cheap Energy Converter

Hutchison was undecided about what to do. He had moved beyond the Hutchison effect and into the field of space energy, and had acquired a Canadian business manager. The winter before his 1995 Japanese tour, Hutchison built a working space energy device about the size of a microwave oven. The Hutchison Converter was based on Tesla's resonance principle. Tesla demonstrated this principle by steadily pulsing bursts of energy into his electric coils, each burst coming before energy from the previous burst had time to die away. This led to higher and higher amounts of energy, like a child going higher and higher on a swing.

Hutchison captured the same pulsing, rhythmic energy by using crystals of barium titanate, a material that can capture the pulses of certain electromagnetic frequencies in the way that a radio can pick up certain radio frequencies. When the crystal pulses, or resonates, it produces electric power.

I saw a demonstration in which the converter put out six watts, enough to power a motor that kept a small propeller spinning furiously. The whirring of a tiny propeller looked rather silly, until one realized that the apparatus contained no batteries, no fuel, and no connection to a power outlet. It worked continuously for months.

One day while experimenting, however, Hutchison cracked a crucial part and decided to take the unit apart.

He built a smaller, more portable model to take on his speaking tour. Resembling an Oscar statue in size and shape, the portable converter put out slightly more than a watt of power. It lit a tiny lamp as a demonstration and also ran a small motor.

At the end of the tour, in front of an audience of about 500 Hiroshima residents, Hutchison slapped the device onto a table lit by the bright lights of a television crew. He quickly unscrewed all the parts and revealed its inner details, while the camera zoomed in for a closeup and a pair of chopsticks provided a scale to show the size of the device. It was clear that the converter contained no batteries. Afterward, men crowded around Hutchison, offering him their business cards and asking him to sell them a supply of barium titanate.

Back home, Hutchison's business advisor fretted that the inventor had given away his secrets. But Hutchison shrugged his shoulders; he had gone beyond the prototype technology he had taken to Japan. He now had a new secret—the stovetop process he called Dirt Cheap because the ingredients included common rocks.

The new process grew out of his use of barium titanate. He wondered, "Why can't I make a material that works even better?" Hutchison knew that other researchers had put electrodes on certain rocks to show that the rocks generated a tiny electric current, somehow soaked up from the cosmos.

So Hutchison sorted through small stones on the street in front of his apartment and threw them into a test tube-sized metal container. Next, he added a mixture of low-cost, common chemicals—he won't reveal which ones—and put this rock soup on the stove to simmer. This allowed water to evaporate and tiny pockets of air to rise from the stones so that the chemicals could enter them. Before the mixture cooled into a solid, he added specially treated posts to draw electricity from the crystal-like substance that had formed. Again, no one is entirely sure as to how the Dirt Cheap method works, although one physicist told Hutchison that the Casimir effect, used by Ken Shoulders to create charge clusters, may be at work (see page 61).

When he first discovered his Dirt Cheap process, Hutchison didn't bother to patent it. He had heard from other inventors how their laboratories had been vandalized and their property had been stolen once the Patent Office had been notified, and he was not eager to be the first inventor to take a bold step by manufacturing a large home- or factory-sized unit that could restructure industries. Besides, in the 1980s—when he was still working with the Hutchison effect—he had received a few threatening comments from strangers.

How could Hutchison enjoy his peaceful life and still get a space-energy product to the public in a low-key manner? He says he has hit upon an unusual strategy: building miniature flying saucers powered by Dirt Cheap-supplied electricity, and selling them as space-energy children's toys. Hutchison hopes an environmentally safe toy that lights up without batteries will intrigue the public into buying Dirt Cheap devices that could power large appliances. And perhaps, the Dirt Cheap process could help lead to a world of non-polluting new energy.

In the next chapter, we will meet an inventor who used magnets to tap the energy of space.

6
Floyd Sweet—
Solid-State Magnet Pioneer

There is suppression launched against any free-energy inventor who succeeds or is very close to succeeding.

—Ret. Lt. Col. Thomas Bearden

Let us all use the knowledge wisely.

—Floyd Sweet and Thomas Bearden

The late Floyd "Sparky" Sweet created a breakthrough magnetic solid-state energy generator. For complex reasons, he did not develop his device into a commercially viable product. However, as a magnetics specialist with a distinguished industrial career, Sweet was not a man whose technical claims could be easily dismissed by critics.

Sweet's story is important for three reasons. First, creditable witnesses saw his invention convert the invisible energy of space into useable amounts of electric power—without fuel, batteries, or connection to an outlet. Second, he was subjected to the same kinds of harassment that the inventors we met in Part I had to face, including threats on his life. Third, and most important, Sweet's research has inspired the work of other space-energy inventors, some of whom may well produce a useful stationary-magnet device.

FLOYD SWEET AND MAGNETS

Floyd Sweet (1912–1995) grew up in Connecticut, in an era when radios were home-built crystal sets. At the age of nine, his intense interest in how things work was directed into building and disassembling radios and other electrical apparatus, such as a small Tesla coil (see Chapter 2) energized by a Model T spark plug.

When Sweet was eighteen, a family friend helped him find work at the nearby General Electric plant while he went to college. He got the nickname "Sparky" after he misconnected some wires one day, which resulted in an instrument exploding in a spectacular spray of sparks. Despite this incident, his employers were pleased with his work—especially his intuitive gift for coming up with answers to electrical problems.

Sweet stayed with GE after completing his education. He worked in the company's Schenectady, New York, research and development center from 1957 to 1962—a dream job in which he could use a well-equipped laboratory to follow his hunches on intriguing magnetics projects. That line of research fascinated him. In 1969, he obtained a master's degree from the Massachusetts Institute of Technology.

By the mid-1970s, Sweet and his wife, Rose, had moved to the Los Angeles area to enjoy semiretirement. Besides serving as one of GE's preferred consultants, Sweet designed electric equipment for other customers.

Floyd Sweet was more than a professional scientist who worked with magnets. He had a passion for magnetism, and for the concept that the entire universe is permeated with a magnetic field. Once he fully retired in the early 1980s, he would have happily spent many hours each day building a device that could tap into the energy of that magnetic field. But Rose fell ill, and was an invalid for the last seven years of her life. This demanded Floyd's attention and forced him to dip into their savings. He also had to cope with his own ill health, including a period of near blindness. Despite these problems, he worked on his device when not preparing meals and tending to his wife's needs.

SWEET'S VACUUM TRIODE AMPLIFIER: DEFYING CONVENTION

For decades, new-energy researchers talked about the possibility of treating a magnet so that its magnetic field would continuously shake or vibrate. On rare occasions, Sweet saw this effect, called self-oscillation, occur in electric transformers. He felt it could be coaxed into doing something useful, such as producing energy. Sweet thought that if he could find the precise way to shake or dis-

Floyd "Sparky" Sweet built the Vacuum Triode Amplifier, which used magnets to serve as gate through which space energy could flow. This allowed such energy to be used as a source of power.

turb a magnet's force field, the field would continue to shake by itself. It would be similar to striking a bell and having the bell keep on ringing.

As usual, Sweet—who said his ideas came to him in dreams—turned for inspiration to his expertise in magnets. He knew magnets could be used to produce electricity, as we learned in Chapter 4, and wanted to see if he could get power out of a magnet by something other than the standard induction process. That process involves either moving a magnet past a wire coil—a coil of conductive wire, such as copper—or moving a coil through the field of a magnet. This changing magnetic field causes an electric current to flow in the copper wire.

What Sweet wanted to do was to keep the magnet still and just shake its magnetic field. This shaking, in turn, would create an electric current. One new-energy researcher compares self-oscillation to a leaf on a tree waving in a gentle breeze. While the breeze itself isn't moving back and forth, it sets the leaf into that kind of motion. Sweet thought that if space energy, discussed in Chapter 4, could be captured to serve as the breeze, then the magnetic field would serve as the leaf. Sweet would just have to supply a small amount of energy to set the magnetic field in motion, and space energy would keep it moving.

By 1985, he had come up with a set of specially conditioned magnets, wound with wires. To test his device, Sweet discharged a current into the wire coil around the magnet. As a result, the coil disturbed the magnet's field. It was as if Sweet had snapped the magnet's field out of position to set it in motion. Sweet then connected a twelve-volt lightbulb—the size used in flashlights—to the coil. If the device was producing electricity, the bulb would light.

The results were more than Sweet expected. A surge of power came out of the coil and there was a bright flash from the bulb, which had received so much power that it melted. Years later, Sweet remembered that Rose had seen the flash and called out, "What did you blow up now?"

The inventor was baffled by the dazzling flash of light—why so much energy? He returned to his workbench to make further models. Needing a theory to explain his startling discovery, he remembered hearing about Thomas Bearden, retired Army officer and nuclear physicist, and John Bedini, an electronics expert, on a local radio show. Sweet called Bedini, who arranged for Bearden to visit Sweet.

Bearden saw the curious device pull nearly six watts of electric power out of the air with only a tiny fraction of a watt going into the machine. Bearden ran tests to his heart's content, and was delighted to see a little unit embodying the unorthodox concepts that he had written about over the years, the concepts behind space energy. He called Sweet's assembly of magnets and wire coils the Vacuum Triode Amplifier (VTA). Bearden decided that the device was serving as a gate through which energy from space was being herded into a electric circuit.

The most amazing aspect of Sweet's device was that it put out so much more power than it took in. How much more? In a 1988 model, Sweet found that 330 microwatts—330 one-millionths of a watt—of input power made it possible for the VTA's wire coils to put out more than 500 watts of usable energy, or about one and a half *million* times the input power.

The VTA's Special Effects and Difficult Development

The VTA turned out to have some very odd effects, but Bearden's research background prepared him for that. So in 1987, Bearden asked Sweet to perform an antigravity experiment. Bearden calculated that the six-pound machine would levitate when about 1,500 watts of power were drawn out of it, but that the magnets might explode at about the same power level. He warned Sweet to limit the output to no more than 1,000 watts. A VTA would be placed on a scale so that its weight could be carefully monitored while it was hooked up to a box of lightbulb sockets. Screwing bulbs into the sockets would draw off the power.

About a week later, Sweet excitedly read off results over the phone to Bearden—who was home in Alabama—as Sweet screwed in ten 100-watt bulbs, one at a time. The device gradually lost weight until it was down to 90 percent of its original weight. For safety reasons, Sweet and Bearden stopped the experiment before the device could begin to hover or fly.

Why did the VTA lose weight? According to Bearden's theory, gravity becomes a pushing force rather than a pulling force under certain conditions. Bearden also says that space energy has a pressure, referred to as energy density. If the pressure above an object is decreased while the pressure under the object is increased, the object will be drawn upwards. The VTA may have changed the energy density by drawing on space energy.

The technology could sometimes do spooky things. Walter Rosenthal of California, a test engineer who has helped many struggling inventors test their devices, recalls an incident that Sweet had told him about. The incident occurred while Sweet was trying to document his antigravity experiment:

> The machine's weight was observed [to be] decreasing with an increased load [of lightbulbs], in a quiet orderly fashion, until a point was suddenly reached when Floyd heard an immense sound, as if he were at the center of a giant whirlwind but without actual air movement. The sound was heard by Rose in another room of their apartment and by others outside the apartment.

This experience has been confirmed by a Canadian space-energy researcher, who heard a similar whirlwind sound during one of his experiments.

Another unusual effect of Sweet's VTA was the fact that it produced cold, instead of the heat usually generated by electric equipment. The inside of the VTA was as much as twenty degrees cooler than the surrounding air. The greater the load put on the device, the cooler it became. When VTA wires were accidentally shorted out, they flashed with a brilliant burst of light, and were found to be covered with frost. One time, a brief contact with the equipment froze some of Sweet's flesh, causing him pain for about two weeks afterward.

Sweet discovered other interesting effects. But development of the VTA was slowed by trouble with materials and processes, and by financial entanglements. Sweet had to find magnets that could hold the self-oscillation effect. That required magnets with force fields that didn't vary much across the face of the magnet.

Also, standard mathematical calculations didn't work with the VTA. In 1991, Sweet produced a math theory for the VTA—an engineering design model that showed how factors such as the number of turns of wire in the coils affected the device's behavior. Producing this theory was an important step. Without it, other researchers would not reproduce Sweet's work.

Sometimes it was difficult for Sweet to reproduce his own work. As with first models of any new technology, the VTAs he built were very unreliable. For example, at times their output went down at night and picked up again during the day. Sometimes, they just plain stopped working for no apparent reason. But when the VTA worked, the power it put out for its size was unprecedented.

Sweet Challenges the Laws of Physics

Bearden contributed to the theory that explained Sweet's invention. Much of the theory that Bearden used to explain how the VTA worked came from advances in the field of phase conjugate optics, a specialized study of light used by laser scientists and weapons researchers. Using information from this field, Bearden said that the VTA was able to amplify the space energy it took in.

The science establishment requires that an invention be explained by accepted laws of physics, and so much output from so little input *seems to* violate those laws, which do not allow for such a thing. However, Sweet and Bearden recognized that these laws apply to ordinary, or closed, systems—systems in which you cannot get more energy out than what you put in. Because the VTA allowed energy to flow in from the vacuum of space, it was not operating in a closed system, but in an open one. (See Chapter 1 for a discussion of closed versus open systems.) A VTA operating in the flow of space energy is like a windmill operating in the wind. Both receive excess energy from an outside source. But since neither operates in a closed system, neither violates the laws of physics.

In 1991, a paper by Sweet and Bearden was read at a formal gathering of conventional engineers and physicists in Boston. Neither Bearden nor Sweet were able to attend—Bearden was called away on business, and Sweet was recuperating from heart surgery. Walter Rosenthal went instead. The paper said that the VTA had the signs of being a true negentropy device, or a device that was able to turn random space energy into usable electricity (see Chapter 4).

How did this work? It helps to think of a handful of marbles on a tabletop. You can either roll them all in one direction, or you can scatter them in all directions. If you scatter the marbles into a reflector, the reflector will roll them back to you in an orderly fashion. Although the language they used was quite technical, what Sweet and Bearden basically said is that the VTA was able to take energy "marbles" and keep rolling them back and forth, building energy as they went along.

After Bearden's paper was read, Walter Rosenthal stood up and startled the audience of skeptical engineers: "I have personally seen Floyd Sweet's machine operating. It was running . . . those small motors you saw in the video. It was jump-started with a nine-volt battery. There was no other electrical input required. . . . There was no connection to the power line whatsoever." And, no, there were no moving parts.

Although most of the audience listened politely, it was too much for one engineering professor. He stalked out of the room, saying, "To present such a remark at an engineering conference is the height of irresponsibility! It violates virtually every conceivable concept known to engineers."

SWEET IS THREATENED

Could activity at the Sweet home been secretly watched by strangers? Sweet told the story of a time in the late 1980s when a man accosted him as Sweet was leaving a supermarket. Sweet remembered the man's expensive-looking shoes, and the fact that he was immaculately dressed. But in the stress of the moment, Sweet couldn't focus on much else.

What made the inventor nervous was the photograph that the man held, a photograph showing Sweet at work on his tabletop-model VTA—in the supposed privacy of Sweet's own home. In what Sweet said was a remarkably clear photo, he was sitting in the dining room on the second story of the apartment building where he lived with Rose.

"He walked me all the way to my building, telling me what would happen to me if I didn't stop my research," Sweet recalled. "How they took that picture through my window, I'll never know." As Sweet remembered it, the man claimed to be connected with a conglomerate that did not want the VTA to come onto the market at that time. He told Sweet, "It is not beyond possibilities to take you out of the way."

Sweet said that afterward he called the FBI in Los Angeles. He believed that two agents staked out his house for a couple of weeks, but that nothing came of it.

Around the time of the photo incident, Sweet was getting telephone calls and death threats from strangers. He said there were "people calling at all hours. The police put a tap on my line and over a six-month period, over 480 calls came in from all parts of the United States. But they were from pay stations." Thus, the police could never find the callers.

Early in the VTA's development, someone broke into Sweet's apartment and stole his notes. He then began to code his notes.

Sweet temporarily stopped work on his invention, out of concern for his ill wife. "They must have known I stopped; they didn't torment me any more."

FOLLOWING IN SWEET'S FOOTSTEPS

On July 5, 1995, Floyd Sweet suffered a fatal heart attack at the age of eighty-three. A couple of weeks before his death, Sweet said that the automotive industry was testing his power unit for use in cars, and that they had a unit running for 5,000 hours. He said he was dealing with people at General Motors, but no one has been able to confirm Sweet's claims.

The VTA itself is bogged down in legal problems. But Tom Bearden, who put much of his own time and money into the project, hopes that the VTA can be resurrected so that the world will realize what a pioneer Floyd Sweet was. And despite the confusion surrounding Sweet's affairs at the time of his death, other researchers are continuing this line of research.

Confusion and Secrecy

The automotive industry may not have been the only potential investor that Sweet was dealing with. At the time of his death, there was some confusion concerning the rights to Sweet's hardware and papers, held by Sweet's second wife, Violet. Bearden says that Sweet signed a number of agreements with a number of backers, and that some of these people have claimed rights to the invention. At least two of these investors say they want Sweet's laboratory equipment, inventions, and technical papers to go into a proposed Floyd Sweet Museum so that other researchers could study the technology. Walter Rosenthal is trying to help all parties work towards an agreement.

Despite Bearden's urging, Sweet never had the VTA certified by independent testing. "He feared that his life would be snuffed out immediately if he even attempted such a thing," Bearden says.

Sweet also frustrated his fellow researchers by keeping secret his most important process—how he conditioned the magnets that are at the heart of the VTA. Did he pump the magnets with powerful electromagnetic pulses to shake up their internal structure? He refused to give details, and said it wasn't likely that other researchers would learn his secrets: "The odds against them finding out is like trying to open a safe with 100 dials set from zero to a hundred, without knowing the combination."

Sweet not only feared for his life, but once said he feared that if he described how he made his device work, unscrupulous people would build models without giving him his due. He was also concerned about what would happen if the VTA was widely sold everywhere at once, replacing many other electric devices. "If it all came out at once, the stock market would collapse," he said. "The government doesn't want it." To be fair to Sweet, I would point out that he is not the only inventor who has been uncomfortable in disclosing key aspects of his work.

Other Researchers and the VTA

Other inventors are trying to carry on Sweet's work. The VTA is well-known on computer bulletin boards that list "free energy" as a topic of discussion. Experimenters scramble for details of how the device was built.

One researcher who has claimed some success is Don Watson, a self-educated inventor from Texas. Watson says he has built a working device similar to Sweet's VTA, which he works on at night after working during the day as a telephone systems installer.

In Somerset, England, electronics expert Michael Watson (no relation to Don) built a replica of Sweet's VTA, but claimed no success in the experiment. Despite that, he says, "In my opinion the inventor of the VTA, Floyd Sweet, has made a scientific discovery of [the] greatest importance."

Watson thinks that attempts to reproduce Sweet's results may run into problems because the type of magnets Sweet used are no longer available. But he says, "The important point about the VTA is that a form of magnetic instability exists that can act as a significant energy source."

When this fledgling space-energy science reaches maturity, what could the VTA do for the lives of the rest of us? Bearden speculates that the new physics will change our lives in undreamed-of ways:

> By mastering, controlling, and gating the vast, incredible energy of the seething vacuum [of space], we can power our automobiles, flying machines, and technology inexhaustibly. Further, it can be done absolutely cleanly; there are no noxious chemical pollutants.
>
> With practical antigravity, ships can be developed to cross the solar system as readily as one crosses the ocean today. . . . The inexhaustible vacuum fills every system, everywhere, to overflowing.

Despite the difficulties that Sweet ran into in his attempts to perfect his invention, he helped science take a leap into the future. It perhaps could have leaped further if he had cooperated more freely with other researchers in the last decade of his life, and if he had been tidier in his business dealings. But Sparky Sweet deserves praise for charting a new course.

In the next chapter, we will meet other energy innovators who have discovered the power of magnets in motion.

7
Rotating-Magnet Energy Innovators

I think it is possible to utilize magnetism as an energy-source. But we science idiots cannot do that; this has to come from the outside.

—Werner Heisenberg,
Nobel laureate

The magnet is a window to the free space energy of the Universe.

—Bruce DePalma,
Inventor

As we saw in Chapter 6, magnets can be used to capture space energy and put it to work. Magnetic fields can be tinkered with so that they serve as gates, guiding space energy into electric devices in the same way that a sluice in a river guides water into a waterwheel. This opens a whole new world of energy possibilities.

This chapter introduces us to two inventors who have shown that it is possible to use magnetism as a power source. Unlike Floyd Sweet and his stationary-magnet device, these men use rotating magnets to convert space energy into electricity. One began his career as a physics teacher at the Massachusetts Institute of Technology (MIT) and is now self-exiled in New Zealand, while the other is an aviation safety consultant who recently gave a lecture to a group of physicists at MIT. We will also see how space energy is being pursued in Asia, even as it is being ignored in North America.

BRUCE DePALMA AND THE N-MACHINE

While his brother Brian has spun a Hollywood career directing films such as *Carrie, Scarface,* and *The Untouchables,* it looked like Bruce DePalma would live a secure life in academia, wrapped in the respect accorded an MIT faculty member. After receiving an electrical engineering degree from MIT in 1958, he worked in both government and industry before going to Harvard in 1961 for graduate work in applied physics. He became an MIT lecturer in the late 1960s.

During that turbulent time, DePalma's life underwent a change, a period of soul-searching that was spurred by both the student movement and by his sense that society was disintegrating. As a result, he dropped out of academia and headed west to Mendocino, California, where he took up meditation. One afternoon, his thoughts turned to something he had played with as a kid and never understood—why does a gyroscope behave as it does? A thought came out of the blue—maybe the rotation of the gyro wheel somehow locked onto the space around a spinning body such as the earth.

Experiments With Rotation and Energy

Sometimes the simplest of experiments leads to new understanding. In the sixteenth century, Galileo's first breakthrough came from dropping a big rock and a small rock from the Leaning Tower in Pisa and finding, contrary to accepted belief at the time, that they both fell at the same rate.

DePalma's breakthough also came from a simple experiment. He rotated ball bearings—steel balls like those found in pinball machines—at a high rate of speed, and launched them into the air while carefully taking multiple time-lapse photographs. He discovered to his surprise that they rose farther and fell faster than ball bearings that were not spinning when launched. He thought this indicated that the spinning bearings were interacting with a new kind of energy—what we now call space energy.

DePalma was even more intrigued when he launched pairs of ball bearings, one spinning to the left and the other spinning to the right. He found that each bearing rose and fell at a different rate, indicating that each might be interacting with this different source of energy in a different way.

DePalma felt his findings were important, and took them to a prestigious mentor of his, a Princeton physicist. But he failed to raise the man's interest.

So DePalma retreated with some friends to a farm in Pennsylvania for more research with rotating objects. Starting with what was at hand, he put the pendulum from a grandfather clock into a vacuum—to rule out any air-pressure effects—and found that spinning the bob did in fact make a difference in the length of the pendulum's swing. He then did an experiment which showed that if you collide a rotating object into something else, it rebounds further than if it had not been rotating. As with the ball-bearing experiments, these results indicated that an object might pick up space energy while spinning. (See "Spirals of Energy" on page 13.)

As a result of his experiments—and of the experiments of others—DePalma now imagines that space energy flows through a metal conductor and gives it different properties, just as fluid flowing into a dry sponge gives weight to the sponge. (See Chapter 4 for a more complete discussion of the theory behind space energy.)

DePalma continued his gravity and inertia research when he moved to a home in the foothills of Santa Barbara, California. His living room was full of unusual sights, such as a circle of grass growing above a spinning stereo turntable and weights hanging from ceiling hooks for pendulum experiments.

DePalma Develops the N-Machine

DePalma decided to take the results of his newfound knowledge from the realm of swinging objects into the realm of electric meters, where accurate measuring instruments are available to everyone. His intuition led him, step by step, to learning about the properties of rotating magnets, and to an energy discovery that further changed his life.

DePalma turned to the writings of the famous British pioneer of electricity and magnetism, Michael Faraday (1791–1867). Faraday is well known for inventing the two-piece induction generator, a piece of equipment that, in its basic principles, is still used to generate electricity today.

But Faraday also invented what he called a homopolar generator in 1831. He found that electric current can be taken from a spinning copper disk when the disk is rotated along with the magnets, instead of past the magnets, as in the induction generator. This unique setup may have allowed Faraday to tap into a different source of energy—space energy. However, Faraday never fully developed the homopolar generator into a fully functional piece of practical equipment. DePalma studied this generator with intense interest, convinced he had found something of tremendous value.

Nearly 150 years later, DePalma repeated Faraday's experiment, except that DePalma used modern materials, such as super-powerful magnets, to extract the electricity. DePalma has named his device the N-machine, "meaning to the nth degree," because he sees the N-machine's potential as being almost unlimited. The name also refers to his speculation that a magnet taps energy from another dimension. He believes the magnets cause a distortion of the aether, a concept we discussed in Chapter 4, allowing space energy to flow into the machine.

From 1978 through 1979, Bruce DePalma and his assistants used the workshop of a California commune—the Sunburst spiritual and agricultural community near Santa Barbara—to build a prototype generator called the Sunburst homopolar generator. After a year of refinements, they began serious testing in 1980. Sunburst test results indicated that output power was more than the input

Dr. Bruce DePalma uses powerful magnets in his N-machine, which puts space energy to work on earth.

power, and that the N-machine was much more efficient than a standard generator.

Then a professor of electrical engineering from Stanford University tested it. Robert Kincheloe did a series of tests on a machine designed by DePalma and built by Charya Bernard of the Sunburst Community from 1985 through 1986. Kincheloe also got more output power than input power. He reported:

> DePalma may have been right, in that there is indeed a situation here whereby energy is being obtained from a previously unknown and unexplained source. This is a conclusion that most scientists and engineers would reject out of hand as being a violation of the accepted laws of physics, and if true has incredible implications.

DePalma Runs Into Trouble

"I thought everybody would beat a path to my door after I did these experiments, but I ran into a stone wall," says DePalma. "It's as if science were in its old age and it's gotten a long way from the laboratory." He adds that it is as if the science establishment took the experiments that were done in the nineteenth and early twentieth centuries, reduced them all to mathematical equations, and made them into a gospel. "If you go to Washington, D.C. to the Department of Energy with a new way of liberating energy, they will bring out all these old relationships and say, `It isn't in accord with the [law of] Conservation of Energy' or `It violates Einstein's Theory of Relativity.'"

DePalma himself had fully believed in the law of energy conservation, which says that you can't get more energy out of a system than you put into it. But what about the results of his experiments? Like most other energy researchers we have met so far, it dawned on him that the excess energy was coming right out of space itself. Therefore, the law of conservation wasn't really being broken.

A skeptical science establishment has not been DePalma's only source of trouble. In 1990, he wrote:

> Three or four commercial groups have approached me to supply money for the commercial manufacture of N-machines. Many promises have been made, but no funds yet. What gener-

ally gums things up is the greed of the money people, not the ability of my machine to perform. . . . What is needed now is a movement to develop the N-machine source of electrical power as a national priority.

At that time, I asked DePalma why he didn't close the loop—feed part of the power output back into the machine to produce continuous motion. Powering a house or a set of appliances with such a setup would be the demonstration that would convince skeptics.

He replied that one reason he hadn't developed the prototype further in the United States was "because I would get my head blown off." He added that a threat was passed on to him through a messenger with highly placed connections to the United States government. In 1992, he perceived that space energy was wanted elsewhere, but not in the United States. Therefore, he expatriated himself, first to Australia and then to New Zealand, where he continues to work on his invention.

BERTIL WERJEFELT AND THE MAGNETIC BATTERY-GENERATOR

Bertil Werjefelt sports a Hawaiian suntan because the islands are his adopted home, but he has little time for the beach. Consulting on aviation safety, overseeing a small corporation, and writing technical papers make up only part of his life. Werjefelt has also been working on a magnet-energy device for several decades. A representative of the Sumitomo Corporation who visited Werjefelt's manufacturing facility said that the invention could be "the most important discovery this century."

Werjefelt was educated in his native Sweden and then came to the United States in the early 1960s. He furthered his education in mechanical engineering at both the University of Utah and the University of Hawaii. He now heads a research and development group, Poly Tech USA, that devises safety equipment for airplanes, such as a system that allows pilots to see the flight path and vital instruments regardless of how much smoke is in the cockpit.

A New Device From Old Concepts

In the 1970s, Werjefelt was one of many people who became concerned about the problem of fossil-fuel pollution. So he used his engineering background to create an energy invention—a generator powered by energy extracted from magnetic fields.

Standard generators, which use magnets, are subject to a problem known as magnetic drag. Drag is a residual magnetism that slows the spinning of the rotor, the part that either moves the magnets past an electric coil or the coil past the magnets, depending on the generator's design. Werjefelt improved the standard generator; he added a special spinning system that cancels magnetic drag by counteracting it with the force fields of additional magnets. The result is a generator that puts out more power with the same input.

That raises a question: Where does the excess energy come from? "I don't know," Werjefelt says. "It could be [space] energy, or something we don't even know about."

Werjefelt's experimental models have not yet evolved into the premanufacturing stage—they have only produced more power output than input for several minutes at a time. But results are impressive enough to keep him going. For example, at one point his generator has shown 160 watts input and 450 watts output, or almost triple the power. He believes his crew has solved some of the most troublesome technical problems and that magnetically powered electric generators could be available for everyday use within a few years.

Some onlookers in the new-energy field are as impressed with the scientific paperwork Werjefelt has done as they are with his experimental models. After he came up with the design, Werjefelt realized that he would need to explain the results in order to get a patent. He would also need to convince a skeptical scientific community.

So Werjefelt dug into the physics literature and found evidence to support his claim. He used this evidence in a 1995 lecture at MIT to argue that standard science's teachings on magnetism have been incomplete from the beginning, and that as a result, the scientific community declared early on that it was impossible to use magnetism as an energy source. The other fundamental forces in nature—nuclear physics and gravitation—have been harnessed in the forms of nuclear power plants and hydroelectric dams, but science has been blind to the possibility of using magnetism as a source of power.

In general, though, Werjefelt refuses to become caught up in what he calls "paralysis by analysis." He is more interested in proving that his device works. "Look at it as a quantum leap in the energy field," he says, "like the leap from slide rulers to handheld electric calculators."

Corporate Interest From Japan

In 1990, Werjefelt sent a notice to large corporations such as General Electric and Westinghouse in the United States, Siemens in Europe, and Hitachi and Sumitomo in Japan about his discovery. Most of the replies were, "It is not possible." Others thanked him and said, "Call us when the patent is issued."

It turned out that the Japanese were very interested in magnets and energy. In October 1993, Japanese television aired a program, *The Dream Energy*, in which Japanese scientist Teruhiko Kawai discussed a device similar to Werjefelt's.

Well-funded Japanese research teams have engineered this discovery into reliable units for existing motors. Werjefelt spent two days with an official from Sumitomo and learned that the Japanese motors are running for hours, days, weeks. Japanese industrialists are switching over to the new units, which will use about half as much fossil fuel as existing motors. For example, the television program showed a refrigerator, a vacuum cleaner, and other common appliances with such motors.

Werjefelt, on the other hand, is more interested in producing electricity. He estimates that if power plants are built using his Magnetic Battery-Generator instead of conventional equipment, they could put out fifteen to eighteen times as much electricity.

GOVERNMENT BACKING FOR INVENTORS ELSEWHERE

As we have seen in Bertil Werjefelt's saga, American corporations are generally staying aloof from new-energy developments, while other countries' governments underwrite corporate research in this field. For example, two countries are working on devices similar to Bruce DePalma's N-machine.

Japan Becomes Involved

In Japan, a soft-spoken scientist is getting government help on his variation of the N-machine. Shiuji Inomata, Ph.D., worked at the electrotechnical laboratory of the Ministry of International Trade and Industry (MITI) in Ibaraki, Japan. Inomata's version of the N-machine—named the JPI, after a private research institute—produced a small amount of excess power as a first prototype.

Dr. Shiuji Inomata
works on a machine
called the JPI that
uses magnets to tap
into space energy.

Now retired, Inomata continues to work on the JPI, and is interested in seeing others continue his research. "Politicians and industry are increasingly becoming aware of the new energy breakthrough," he says. This could give Japan a considerable lead in the race to produce N-machine technology. For further discussion on why new energy fascinates the Japanese, see page 101.

India Also Pursues Space Energy

Japan is not the only Asian country that is actively pursuing space energy. In India, a government-employed nuclear scientist is also working on a type of N-machine—with his employer's blessing.

Paramahamsa Tewari, Ph.D., is a senior engineer with the Department of Atomic Energy's Nuclear Power Corporation (NPC). His version of the N-machine is called the Space Power Generator (SPG). Among the Westerners who have encouraged Tewari over the years is Bruce DePalma. Tewari says, "But for DePalma, I wouldn't have been able to tie up my theory. He was working on similar ideas and kept sending his results to me."

Dr. Paramahamsa Tewari of India is in an unusual position: he works at a nuclear plant during working hours and builds his Space Power Generator in his spare time.

Tewari is project director of the NPC's Kaiga Project in the state of Karnataka. Although his spare time for refining the SPG is limited, Tewari is enthusiastic about it. The NPC's managing director, S. L. Kati, says, "Tewari's prototype SPG can be considered a major breakthrough."

It is unusual for a government to encourage one of its nuclear physicists to explore space energy. But Tewari has gotten special treatment from his government. For example, instead of travelling on a private passport to a new-energy symposium in the United States several years ago, Tewari's passport had been cleared by the Indian government, which smoothed his way through airports. In building SPG prototypes, he uses the services of electricians and mechanics, as well as a workshop, at the nuclear plant where he works. Tewari is pleased with how things are going at his day job—the project is moving forward. Thus, he feels well justified in putting a "do not disturb" sign on his door twice a week to work on the SPG for a couple of hours.

Why has Tewari found such cordiality from an agency that provides megaproject power? He says, "They feel that if something meaningful comes out of (the Space Power Generator), the world may benefit." He adds:

I am heading the whole electrical department of a nuclear project. . . . I do my job great, and there is mutual respect. People
didn't [get] in my way. I also very bluntly threw away any
opposition. I just said, "Look. I don't care about you. I earn my
living as a government officer, yes. But I have my research to do
and you can't stop me."

Space energy is not the only new-energy source under development. In the next part, we will look at other incredible technologies.

Part III
Emerging Energy Technologies

I recently read—with sympathy—a letter to a newspaper, written by a young man who lives in a small Canadian town. He says that his generation is being left with a staggering mess. Referring to the binge of natural-resources consumption, such as the squandering of fossil fuels, he basically says that the party is over on this planet.

His bitterness is typical of many young people who look at our environmental problems and despair for their future. They seem to have inherited a planet that is being unnaturally warmed by greenhouse gases and littered with radioactive debris, while its waterways are smeared with oil slicks.

Their pessimism could be lightened, however, if they considered the scope of emerging energy technologies: heat itself as a source of nonpolluting power, water as a fuel. The very variety of answers to these pressing environmental questions is in itself a cause for optimism. Effective tools for planetary cleanup are being developed by unexpected heroes. When people of all ages realize this fact, they could be galvanized into positive action, such as lobbying for sensible national energy policies. Scientist Hal Fox is such an optimist; he joyously describes the future as the Enhanced Energy Age.

In this section, we will look at some of the emerging energy technologies that are close to commercial production.

8
Cold Fusion—
A Better Nuclear
Technology

*We believe that . . . there will be cold fusion powered automobiles,
home heating systems, small compact electrical generating units,
and aerospace applications.*

—Dr. Eugene Mallove and Jed Rothwell,
Cold Fusion magazine

*I would have thought that a field having such potential importance
to energy production and nuclear theory would generate more
curiosity.*

—Edmund Storms,
Radiochemist

I n 1989, scientists Stanley Pons and Martin Fleischmann did
what science said was impossible. They announced that they
had discovered cold fusion—the joining, or fusing, together of
atoms at room temperature accompanied by the release of excess
energy. It had generally been believed that such fusion required
extremely high temperatures, and could take place only in reactors
costing billions of dollars. But Pons and Fleischmann said that they
had achieved fusion in a tabletop device they had built themselves.

The arguments began as soon as the shock wore off. The fact that
Pons and Fleischmann had made their announcement to the press
did not sit well with many scientists, who said that the pair should
have first published their findings in one of the established scien-

tific journals. Others tried unsuccessfully to duplicate the pair's results, and then criticized Pons and Fleischmann for poor experimental technique. And later that year, the United States Department of Energy (DOE) decided to not fund cold-fusion research.

None of this has stopped Pons and Fleischmann, along with a host of other scientists, from pursuing cold fusion. But first, a question: what is fusion?

FUSION HOT AND COLD

Fusion is the opposite of fission, although both processes start with atoms. Atoms are the tiny building blocks that make up all matter. An atom consists of a nucleus, which is made up of protons and neutrons, and electrons, which form a cloud around the nucleus. Different atoms contain different amounts of protons, neutrons, and electrons, and form different types of matter.

Fission is the splitting of an atom's nucleus, such as by bombarding it with neutrons. This releases a great amount of energy. Materials with heavy, unstable atomic nuclei, such as certain kinds of uranium, are needed for fission. An atomic bomb and a nuclear power plant both use fission.

Fusion is the joining together of atomic nuclei. Hot fusion, which is said by some scientists to be what energizes our sun, uses a form of the lightest element, hydrogen.

Textbooks teach that temperatures reaching millions of degrees Fahrenheit are needed before the positively charged hydrogen nuclei can overcome their natural repulsion toward each other, since like charges repel—think of what happens if you attempt to bring the north poles of two magnets together. If the hydrogen nuclei do come close enough together, they form something different—helium nuclei. In the process, tremendous amounts of energy are released. The hydrogen bomb uses hot fusion, but a fusion-based nuclear plant, long a goal of scientists, is decades away.

On the other hand, even those who believe in the existence of cold fusion—the joining of atomic nuclei at normal room temperatures—are not entirely sure of how it works. Instead of using superheated gas, cold fusion seems to be based on the reaction of a metal such as palladium, which has large spaces between its nuclei, and a liquid form of hydrogen called deuterium. The deuterium seems to move into the spaces within the palladium in the same way that water moves into the open, absorbent surface of a towel.

No one disputes the fact that the metal absorbs the deuterium. However, cold-fusion proponents cannot prove that the reaction which follows the absorption is a nuclear reaction. Thus, most conventional scientists reject the whole idea of cold fusion.

The problems of both fission and hot fusion, which include high costs and radioactivity dangers, are well known. Cold fusion, though, is not without its own problems. For example, one of the byproducts of cold fusion is the radioactive gas tritium, a rare form of hydrogen. Thus, cold fusion may pose more of an environmental problem than other energy alternatives. As one new-energy organization has noted, cold fusion introduces concerns about radioactivity, and even a low level of radiation can eventually lead to environmental and health problems.

On the other hand, cold-fusion scientists say that the amount of tritium produced in cold-fusion reactions is only a tiny trace quantity. Tritium has a short half-life—only about twelve years—and can be easily shielded by a thin metal foil, in contrast to the thick layer of concrete needed in a standard nuclear plant.

STANLEY PONS AND MARTIN FLEISCHMANN: MEETING WITH THE PRESS

In the late 1940s, a bright young scientist named Martin Fleischmann was working on his doctorate at Imperial College in England. His thesis was on the topic of how hydrogen diffuses through platinum—a topic that would lead to a firestorm of controversy forty years later.

But meanwhile, Fleischmann earned a reputation as a distinguished electrochemist. Of the countless research projects he knew about, one in particular influenced his thinking. He was aware of projects both in the United States and in the former Soviet Union that used deuterium and palladium under extreme pressures. The rapid rate at which deuterium moved through palladium's internal structure mystified him, and he wondered whether this process could bring about cold fusion.

While working for the University of Southampton he had opportunities to travel to the United States. By the 1980s, he and a colleague in Utah, a younger scientist named Stanley Pons, experimented with Fleischmann's idea of metallic nuclear reactions on their own. Whenever the British scientist was able to spend time in Salt Lake City with Pons, who was head of the University of Utah chemistry department, they worked on an amazingly simple table-top device in order to do the impossible—create room-temperature fusion. To their surprise, they found that the device produced excess heat, more heat than could be explained by either the electricity that went into the process or any known chemical reaction.

Meanwhile, physicist Steven Jones was doing almost the same thing up the road at Brigham Young University. At the time that Jones was about to announce his own work, Pons and Fleischmann were not ready to make an announcement. The duo did not yet have a reliable experiment; sometimes it worked and sometimes it did not. The race to be first involved high stakes, such as patent rights to a world-shaking invention, but they refused to announce prematurely.

As a result of information accidentally given out by others, however, they were forced to do something dramatic. Their controversial press conference of March 23, 1989, was the result. The pair announced that they had measured excess heat in special electrolytic cells. Pons and Fleischmann thought the heat came from nuclear fusion—a cold fusion that could mean a new source of abundant power for homes and industry, and possibly for automobiles as well.

What Is an Electrolytic Cell?

What was the device that Pons and Fleischmann built? Essentially, it was a one-quart mason jar filled with deuterium. A platinum wire and a one-inch-square sheet of palladium attached to a wire were then added to the jar. Normally, passing an electric current through such a system results in electrolysis, in which oxygen and deuterium are released in the form of gas bubbles.

But this time, Pons and Fleischmann found that the cell generated excess heat, more heat than could be accounted for under normal circumstances. They decided that the nuclei of the deuterium atoms were being forced into the atomic structure of the palladium, and were being forced close enough together to create another element, thereby releasing heat.

Rocking the Boat: Cold Fusion Is Attacked

The scientific community was appalled at the way in which Pons and Fleischmann made their announcement. John O'Malley

Bockris, Ph.D., of Texas A and M University, a Pons and Fleischmann supporter, recalled that furor—in a tone of mock horror—while speaking at a new-energy symposium: "That one should announce a super-discovery on the news hour, on the television . . . was the most dreadful thing a scientist can possibly do."

While his audience chuckled at that thought, Bockris noted that Pons and Fleischmann did make a point by puncturing the biggest balloon in science—the funding of hot fusion. The United States has spent about $10 billion over decades of building large fusion reactors that rely on either magnetic fields or lasers to compress and heat fusion fuel. If one counts the salaries of scientists working at national laboratories, hot fusion ate up about half of the budget of the National Science Foundation—a research and education agency in the government's executive branch—in recent years. Meanwhile, Pons and Fleischmann had spent about $100,000 to create their electrolytic cell.

As Bockris put it, "The announcement seemed to make this giant physics establishment look absolutely silly."

But the establishment did respond in 1989: hundreds of physicists who had never handled an electrochemical cell before—and some who had—tried to repeat the experiment. Virtually nothing happened, although a few researchers did get positive results. Therefore, the establishment decided that cold fusion was not worth the bother.

Opponents judge cold fusion in the same terms as hot fusion, ignoring the fact that since cold fusion uses metals instead of gases, it could be a different type of nuclear reaction altogether. For example, critics deride the lack of radiation from cold fusion, as compared with the large amounts of radiation from hot fusion. They say that a nuclear reaction giving off any significant amount of heat would also produce a great deal of radioactivity—enough to kill the experimenter and destroy the apparatus.

There are other things that scientists don't understand about cold fusion: Why are there so many ways—radio frequencies, heat, sound—to trigger it? How can it work with so many types of conducting materials? More importantly, where is the extra energy coming from? Established physics doesn't have the answers. Cold fusion defies established fusion theories, and thus is attacked.

The most remarkable aspect of this story has been the unscientific attitude that has emerged: a blinding hostility from many in the scientific establishment. Apparently, there has been great anger at the fact that outsiders—chemists—are dabbling in what had been previously the territory of only physicists. That hostility has forced some cold-fusion researchers to duck out of sight.

Among the victims of this hostility were Pons and Fleischmann themselves. They were both ridiculed viciously in public, and even accused of fraud. The barrage of criticism came first from fellow scientists. The media quickly joined in, making the duo an object of laughter. In 1993, the Canadian Broadcasting Corporation and the British Broadcasting Corporation produced one of the first programs to present cold fusion in a favorable light. In that program, Stanley Pons's fine character revealed itself. Although he has been the target of perhaps more ill will than the senior scientist who is his friend and partner, Pons talked about the shameful treatment of Martin Fleischmann rather than about what had happened to himself.

Bockris is one of the world-class scientists who does cold-fusion work regardless of the high-level hostility. But even his excellent reputation has not protected him from harassment, such as when a petition was passed around at Texas A and M in late 1993 demanding that his title of distinguished professor be stripped from him. It took an inquiry to clear his name.

The hostility to cold fusion has taken other forms. Being shut out of the standard scientific journals frustrates many researchers in the field. These journals require that a submitted article be judged by scientists working in that journal's field of interest. Articles that don't pass muster are not published. Some researchers have labelled these journals "consensus science publications"—publications that accept only those ideas agreed upon by most scientists. Since cold-fusion experiment results are considered "impossible," any scientist reporting them is presumed to be making mistakes.

Cold fusion has also received short shrift at the United States Patent Office. Patent applications were being turned down, with the Patent Office citing three negative articles: a Massachusetts Institute of Technology (MIT) paper, a *New York Times* article, and a *Washington Post* article. In 1992, *Fusion Facts*, a cold-fusion newsletter, contacted the Patent Office section that handles cold-fusion applications. According to Hal Fox, Ph.D., the newsletter's editor, the section supervisor was asked why patent examiners did not use the many positive articles from such journals such as *Fusion Technology* and *Journal of Electroanalytical Chemistry*. The supervisor replied, "I guess our people don't have access to these publications." However, the Patent Office's head librarian said that such publications were available to patent examiners.

Fox writes, "The conclusion here at *Fusion Facts* was that someone or some group in Washington has a great deal of influence," possibly enough to suppress cold-fusion patents. And cold-fusion defenders have criticized the MIT paper for improper evaluation of data. Despite all of these obstacles, a cold-fusion patent has finally been granted.

Why the big opposition to cold fusion? "A great law we all used to believe in—that nuclear reactions only take place at huge temperatures—is not true," explains Bockris. I believe that it is sometimes very difficult for experts in a given field to have their worldview changed by new, contrary evidence. Therefore, it may be that physicists trained to believe that fusion only occurs under certain conditions simply cannot accept that there may be another way.

DISINTEREST IN AMERICA, INTEREST IN JAPAN—AGAIN

In contrast with the attitude of the Western hot-fusion establishment, Japan calmly treats cold fusion—referred to in that country as new hydrogen energy—like a new branch of physics. In late 1992, Japan's Ministry of International Trade and Industry (MITI) announced that it had set up a center for cold-fusion research with a four-year budget of $25 million, with additional money from about fifteen Japanese companies interested in pursuing this research.

Another example of Japan's interest in cold fusion is the way the country has supported the work of Pons and Fleischmann. In early 1990, the pair gladly accepted a dream offer from a company linked to the Toyota Motor Company—the chance to work in a multimillion-dollar laboratory in southern France, their location of choice.

This is not the first time that Japan has welcomed technology that has been attacked in the United States—see the N-machine story in Chapter 7. Why is Japan so open to these ideas? There appear to be several factors:

• *Need*. Unlike the United States, Japan must import all of its oil. This gives Japan a strong incentive to find new energy sources.

• *Risk tolerance*. Japan's need for new energy sources leads the country to take greater risks in developing such sources. One MITI official describes this attitude as "technological optimism."

• *Deference.* Not all Japanese scientists believe cold fusion will work—one actually said that he would quit his job, shave his head, and become a Buddhist monk if it did. But unlike Western society, which is based on confrontation, Japanese society is based on the avoidance of confrontation. Thus, cold-fusion researchers in Japan do not face the sort of open hostility that they do here.

• *Open-mindedness.* Some researchers speculate that space energy enters into the cold-fusion process, and that is what accounts for the excess heat which is generated. Why would the Japanese accept that possibility more readily than many Westerners do? Shiuji Inomata, the scientist we met in Chapter 7, says that many Easterners believe in an all-pervading energy throughout the universe that can be used for healing. He says this belief allows people to accept the existence of this energy under other names.

In contrast to the Japanese government, the United States government has given only token funding to cold-fusion research while spending vast sums on hot fusion. In 1989, the Energy Research Advisory Board (ERAB), which advises the DOE on which science projects should be funded, wrote a report critical of cold fusion. This report was the justification used for shutting the public purse on cold-fusion work in the United States. The ERAB report was described by one scientist at the national laboratory in Los Alamos, New Mexico, as "a very incomplete and harmful document showing only a minor amount of objectivity." Some Los Alamos scientists who did do some cold-fusion work had to stop because money for the experiments was withdrawn after the results were published.

And not only is the government not funding cold fusion, it is not even keeping its own researchers abreast of what is happening in the field. As a retired Los Alamos scientist, Edmund Storms, Ph.D., puts it, "When the DOE attitude eventually changes, as it must, the [Los Alamos] staff will not even be able to write useful proposals."

WORK ON COLD FUSION CONTINUES

Amid the controversy, work on cold fusion continues in thirty countries. Some of it is being done quietly. As one cold-fusion writer says, "We know of weapons-related work; we know of commercially secret work; but we probably have heard about only a fraction of either." And even some of cold fusion's opponents are yielding to the barrage of scientific output. For example, Storms was finally able to get a favorable article published in Massachusetts Institute of Technology's *Technology Review*.

In North America, cold-fusion work continues in both universities and about a dozen private corporations. According to cold-fusion researcher Eugene Mallove, some of the world's largest corporations are preparing to jump into cold-fusion research.

New ideas are quickly propelling cold fusion toward the marketplace. One example is a cold-fusion cell invented by James Patterson, Ph.D., of Dallas. It is what scientists call a robust piece of equipment, running undisturbed through four days of poking and prodding at a new-energy conference. The cell, manufactured by Patterson's company, Clean Energy Technologies, Inc., puts out much more energy in heat than it takes in as electricity.

Scientists are working on a number of other ideas:

• Other substances, such as regular water, molten salts, and bronze crystals, may be as useful as deuterium in creating cold fusion.

• It may be possible to achieve fusion in tiny imploding bubbles, produced by bombarding a liquid with sound waves. The implosions create pressure, which in turn leads to fusion.

• If the particles from a substance like deuterium vibrate in unison while trapped within a metal's structure, they may tap into space energy (see Chapter 4), which may somehow account for the excess heat.

Of all the new-energy ideas, cold fusion may be the closest to commercial development.

In the next chapter, we will meet two of the inventors who have tried to introduce the world to a hydrogen-based economy.

9
Powering Up
on Hydrogen

The only reason we're years away [from alternatively powered cars] is political. It's not scientific.

—John O'Malley Bockris,
Physicist

My role in the emerging hydrogen economy [is] implementation of technologies which so many people had previously declared to be impossible.

—Roger Billings,
Inventor

Countries and individuals alike are affected by the prices of gasoline, coal, and oil. Sometimes, individuals have to fight faraway wars for the sake of a country's oil-based economy.

The twentieth century has seen a fossil-fuel economy become entrenched. However, like the dinosaurs, it is doomed to die. Fossil-fuel supplies are limited, as is the planet's ability to absorb fossil-fuel pollution.

Futurists have long predicted that humankind will switch to a hydrogen economy. In contrast to the oil economy, a hydrogen economy would be built on a pollution-free, plentiful product. Hydrogen can be obtained from water. Thus, even if a country had a shortage of lakes and rivers, it would never need to fight for energy supplies as long as it touched an ocean.

There are innovators who are working to make a hydrogen economy happen soon. In this chapter, we will discuss who would control a hydrogen economy. Then we will meet Roger Billings and the late Francisco Pacheco. With little help beyond the moral support of their families, these men kept working towards the goal of a hydrogen economy. They are only two of many hydrogen innovators.

WHO WOULD CONTROL A HYDROGEN ECONOMY?

Although it is a conventional alternative fuel source—as opposed to space energy (Part II) or cold fusion (Chapter 8)—hydrogen is included in this book because the leading edge of independent hydrogen researchers are working on revolutionary, small-is-beautiful technologies, technologies that could allow homeowners and industries to get off the electric power grid.

Hydrogen, the lightest substance known, fuels our sun. It burns at a higher temperature than gasoline and contains much more energy than a similar volume of gasoline. Hydrogen—a colorless, odorless, tasteless gas—combines readily with other elements, and thus is rarely found alone in nature.

Despite its bland qualities, hydrogen could dethrone King Oil. While use of fossil fuels spews carbon dioxide and poisonous carbon byproducts into our environment, hydrogen is a carbon-free fuel. The product of properly burned hydrogen is water vapor. Thus, a hydrogen-based energy technology could clean up our environment. (If you're concerned about hydrogen's safety, see page 108.)

Hydrogen has other advantages over oil. As the ninth most abundant element on earth, the supply of hydrogen is virtually endless. Water—from which hydrogen can be obtained—is nearly everywhere on this planet, as opposed to the limited reserves of oil or pockets of coal and uranium. The water in our oceans, lakes, and rivers could supply all the hydrogen we need, if an efficient process could be used to split the hydrogen in water from its bonds with oxygen. An electric process called electrolysis usually does this splitting, and that process could be powered by using another source of energy, such as the sun or the wind.

Some scientists have worked out details for a hydrogen economy in which hydrogen would be stored in tanks and transported by pipe or truck to be sold as clean, abundant fuel. It would be relatively easy for a developed country to switch to hydrogen on this basis without major shifts in the current system of ownership, in which most energy assets are owned by a small number of large corporations.

But other researchers have a more revolutionary vision. They have figured out how to generate hydrogen *at the point of use*. As with the other revolutionary energy alternatives, this would allow for *decentralized* energy sources—individuals and businesses generating their own power.

Researchers have told government officials about these emerging technologies, but government reports to the public on energy do not include these revolutionary water-as-fuel advances. I believe that some decision-makers in government are worried about losing revenue from gasoline taxes, since it would be difficult for the government to collect a fuel tax if you could fill your tank from a garden hose. Perhaps as a result, these problems are not being debated in public forums.

Powering up on one's own is not something new: consider self-taught inventor John Lorenzen of Iowa. When a utility company strung the first power lines into his region in 1940, Lorenzen refused to hook up—he didn't want to pay the service charge of three dollars a month. Lorenzen's ninety-eight acres of corn, now farmed by his son, are dominated by two thirty-foot-high windmills. The energy they generate is stored in dozens of batteries and then converted into standard household current. A reporter found him converting his truck to run entirely on hydrogen, with plans to also convert his household appliances to hydrogen.

Power monopolists seem to have a different scenario in mind when they praise hydrogen, and it doesn't include encouraging individuals to generate their own fuel. Many books that promote a solar-hydrogen economy speak of using a system of pipelines and trucks to deliver hydrogen. This would not change our lives much. It seems that they see hydrogen as a product that can be controlled and parcelled out to the consumer, just as oil and gas are today.

FRANCISCO PACHECO: A DREAM FRUSTRATED

When Francisco Pacheco was a young man in Bolivia on his way to becoming a self-taught scientist, he tinkered in a makeshift laboratory. One day, a match lit for a cigarette changed his life when it touched off bubbles of hydrogen that were forming in a beaker. The bubbles exploded, blowing a hole in Pacheco's ceiling. This led him to marvel at the tremendous power of hydrogen, and he turned his talents to making a hydrogen generator.

More than thirty years later, a Pacheco generator powered a his-

Hydrogen, the Hindenburg, and Safety

Unfortunately, many people associate the word "hydrogen" with one of two things: either the hydrogen bomb or the Hindenburg. Creating a hydrogen bomb involves nuclear fusion, and requires an atomic bomb as a triggering mechanism—that's how much energy is needed to start a conventional fusion reaction. This amount of energy would be impossible to produce (even if it were desirable!) in a hydrogen-powered car or home.

Many people have seen the dramatic photographs and newsreel footage of the Hindenburg fire. The Hindenburg was a German passenger airship that caught fire in 1937 at Lakehurst, New Jersey, with nearly 100 people aboard. Although there were thirty-six deaths, most of those who died jumped or fell, while many of those who were trapped within the ship were crew members at their posts. Hydrogen was used in the balloon itself as a lifting gas, but hydrogen flames generate little heat—compared with the heat-retaining diesel-fuel flames, which burned for three hours—and hydrogen rises as it escapes.

This tragedy has dampened enthusiasm for the use of hydrogen as a fuel ever since.

---

toric boat ride. On July 27, 1974, at Point Pleasant, New Jersey, Pacheco ran a twenty-six-foot boat for nine hours using seawater as fuel. His invention separated hydrogen from seawater as it was needed, and then put the hydrogen's energy to use. The implications were tremendous—an ocean-sized fuel tank full of free energy. Instead of the usual sheen of pollutants from a motor's exhaust covering the ocean, the generator's "waste" product was clean water.

So why did Pacheco die without widespread recognition? After all, according to journalist Karin Westdyk of New Jersey, who saw

Pacheco's generator at work and interviewed him before his death in 1992, Pacheco built prototype generators that fueled a number of machines. But the young man with great dreams in Bolivia ran into indifference and misunderstanding in the United States.

A Hydrogen Generator

Francisco Pacheco (1914–1992) was a short, soft-spoken man who loved children and animals. He could not afford advanced schooling as a young man in Bolivia, but he had an avid interest in the natural sciences. He always appreciated the mysteries of nature, even when fighting with the impoverished Bolivian army in the 1930s during a war with Paraguay over the Chaco border region.

After leaving the army, Pacheco became interested in things electrical, especially in batteries. But after the hydrogen-bubble explosion in his laboratory, he turned his attention to the challenge of extracting hydrogen from salt water on demand. It was a dream he pursued for fifty years.

Pacheco's generator used plates made of two different metals within a stainless steel container. In a process that's not clearly understood, he was able to generate electricity within the generator, and then use that electricity to break up seawater into hydrogen and oxygen.

Pacheco used his generator to run a car, a motorcycle, a lawn mower, and a torch, as well as the boat. His daughter, Irene, says that the boat ran smoother on hydrogen from her father's generator than on the fossil fuel it was designed to use.

A Promise Unfulfilled

Pacheco's plans for his invention started out well enough. In 1943, United States Vice-President Henry Wallace, who was making a goodwill tour of South America, witnessed Pacheco's generator as it powered an automobile. Wallace invited him to bring the generator to Washington where, later that year, he demonstrated it to scientists and representatives from the War Department at the Bureau of Standards. He applied for a patent, but because the United States was at war, all patents were sealed and available only to the military.

At any rate, energy costs were not an issue, since oil was cheap in those years. Pacheco's lawyer advised him to set aside the patent application. So Pacheco applied for United States citizenship, brought his family to his adopted country, and waited for the right time for alternate energy technology to be appreciated. He found work in defense plants during World War II, and then as a heating engineer in New York City until his retirement in the late 1970s.

During the Middle Eastern oil embargo in the mid-1970s, Pacheco decided to see if anyone would be interested in his generator. But he soon concluded that neither the energy industries, the heavily subsidized utilities, nor the Department of Energy (DOE) were interested in developing clean, abundant, safe energy from hydrogen.

There was no lack of effort on Pacheco's part. In 1975, his letter to the DOE was returned with a request for more details. When Pacheco provided them, the DOE referred him to the Bureau of Standards, stating that the material was not organized enough to give to their technicians. In 1986, Pacheco contacted the DOE again, and received a form letter with a summation of the virtues and drawbacks of hydrogen as a fuel. One of the drawbacks referred to the problem of storage, despite the fact that Pacheco's generator created hydrogen on demand, and therefore didn't require any storage capacity.

Pacheco patiently wrote back, explaining that his system produced hydrogen only when needed, so with his system there would be no need to store the hydrogen. "His detailed response was ignored," says Westdyk.

In the meantime, a young engineer learned of Pacheco's generator and became very excited. He said he could not wait to tell Con Edison, New York City's utility company, about it. But Con Ed didn't even want to hear about the generator.

Pacheco's need to provide for his family meant that he was never able to pursue the higher education he always wanted. Therefore, in an effort to overcome the skepticism he faced because of the "Ph.D." he could not add to his name, Pacheco had his invention analyzed by several independent laboratories in 1979. It passed all tests.

He sent letters to automobile companies. No interest. He sent letters to oil companies. The same—one oil company engineer said that it would be against the company's interest to develop Pacheco's system. He sent letters to about thirty utility companies. No response. He sent letters to all 100 United States senators. Only two of whom responded, and nothing came of those contacts.

Like many other inventors who have run into this wall of silence but have refused to give up, Pacheco built demonstration models to show government and industry officials. His important visitors admitted they were impressed, and promised to help. But none ever did. For example, in 1974, Pacheco showed his generator to his Congressional representative, Robert Roe. The legislator promised to tell Washington officials about the clean-energy invention, but Pacheco never heard from him again.

Pacheco still did not give up, and in 1977 his prototype system energized his neighbor's 1,000-square foot home in West Milford, New Jersey. Pacheco paid for the system by borrowing money against his own home. It provided electric energy as well as hydrogen fuel for cooking and heating. New Jersey's energy commissioner visited with several members of his staff. They were impressed, and wrote a letter to the DOE. Again, nothing happened.

The persistent inventor then tried the media. Television talk show host Geraldo Rivera expressed interest after the 1974 powerboat demonstration, and wanted to do a show about the generator. The idea was turned down by the television station involved.

In 1980, after the Inventor's Club of America inducted Pacheco into their Hall of Fame, the television news show *60 Minutes* contacted him. The TV crew visited the Pachecos at their West Milford home, and filmed the generator as it made hydrogen fuel for a Bunsen burner and for a torch that sliced through a three-quarter-inch steel plate. There were other visual demonstrations, such as hydrogen inflating a balloon and running an electric motor.

One of the demonstrations involved a lawn mower, and did not go as well as it usually did. Pacheco had rushed out to buy a new lawn mower for the filming, and did not have the time to test it out. As a result, the engine choked due to the excessive amount of hydrogen being produced. But this did not seem important in light of the success of the burner, the torch, and the motor. The *60 Minutes* crew reassured him that they had enough material to present an entire show with the successful demonstrations.

Westdyk writes, "Later when the show was aired, Pacheco was devastated, as the show had a completely different focus. The only demonstration aired was the lawn mower, and it was used to provide an example of an independent inventor's non-working invention."

Pacheco demonstrated his generator at several new-energy conferences and meetings. After getting compliments from scientists at these meetings, he applied for another patent. In 1990, U.S. Patent No. 5,089,107 was issued in his name for the Pacheco Bi-Polar Autoelectrolytic Hydrogen Generator. However, before he could try to put his ideas to use, Pacheco died. To the end, he had tried to interest the world in his invention, but had gotten nowhere.

This does not mean that the Pacheco generator is finished. Several people—including Pacheco's grandson, Edmundo—have written about the generator, and there are parties that are interested in developing the technology into a marketable product.

ROGER BILLINGS: RACING AGAINST TIME

A hydrogen innovator who has had more success in developing his inventions is Utah native Roger Billings. But his career—and his younger brother's life—nearly ended when Billings was only fifteen years old. In 1963, he took the engine out of a gas-powered lawn mower, removed the carburetor, and installed his own invention—a glass flask partly full of water, to which various fittings and tubes were attached. The persuasive youth convinced his mother to sign for purchase of a hydrogen cylinder, and he then taught his brother, Lewis, how to operate the cylinder's valve.

At this point, religious training saved the day. Roger decided that a short word of prayer would be in order first. As he and his brother prayed, Roger worried a bit about the large glass flask. In the garage, he found his father's heavy flying jacket. The boys zipped it around the flask and tied it down.

After dramatically counting down from ten, Roger pulled the starter rope with all his strength. The engine sounded like it was going to catch, but then a spark got out and his perfect mixture of hydrogen and oxygen ignited—in the tubing, not the flask. This was not supposed to happen. Fire shot back into the glass flask and created a terrible explosion. Mother came running outside. The aviation jacket was destroyed, and all that remained of the flask was the neck. But the boys were not scratched.

The career that got off to such a bang did not end there. Roger Billings grew up to be a still-enthusiastic, articulate man with a magnetic personality. He has devoted his life to assembling a hydrogen-energy technology that could win a race against time, the race Billings refers to as "our race . . . to keep this planet a nice place to live."

Billings and the Push for Cheap, Safe Hydrogen

Billings's main contributions to a workable hydrogen technology are in the areas of price and safety.

His search for an efficient hydrogen technology started with known principles. Many a high school physics teacher introduces students to the equation "hydrogen and oxygen yields water and energy" by showing how a flaming match ignites hydrogen bubbling out of a laboratory flask in popping mini-explosions. When hydrogen burns in air, it creates water vapor, which can then be distilled into water and electrolyzed—broken down into oxygen and more hydrogen.

Billings learned how to incorporate this burn, distill, burn cycle into safe, cost-effective systems for the home and the car. His technology makes it possible to overcome a barrier to the use of hydrogen-powered vehicles—the lack of an infrastructure, such as the system of pipelines, trucks, and service stations through which gasoline is distributed. His fuel cell can be run in reverse to generate hydrogen, allowing the homeowner or car owner to stockpile hydrogen for future use.

Dr. Roger Billings of Independence, Missouri, has developed innovative ways to use hydrogen in vehicles and homes.

Hydrogen has always posed another problem—safe storage. Billings eventually found that metal hydrides are the answer. Metal hydrides are combinations of metals that soak up hydrogen like a sponge under certain temperature and pressure conditions and turn the hydrogen into a powder. When the pressure is reduced—such as by opening a valve—and heat is applied, the hydrogen is released.

Although Daimler-Benz, the company that makes the Mercedes, has been widely credited with pioneering the use of hydrides, Billings says that the company learned about this method of hydrogen storage from him. He says that German engineers visited his Utah laboratory, learned the technology, and went home to build their own prototype.

Billings tells about tests in which a metal hydride tank, fully charged with hydrogen, was thrown into a bonfire without exploding: "The army came and shot armor-piercing incendiary bullets through it and still couldn't get it to explode." This means that the average person can use this type of hydrogen-powered device without fear of explosion or fire.

Taking on the Business World

Roger Billings learned as a young man that not every institution opens its doors to an innovative researcher. In 1972, he was working at Brigham Young University on a research grant from the Ford Motor Company. His aim was to learn how to get rid of the trace amounts of nitric oxide that pollute the otherwise pure exhaust from a hydrogen-fueled car. His method of injecting droplets of water into the combustion chamber seemed to work, according to computer simulations. But there were no grants to try it out.

Despite the lack of hard evidence, Billings and his team of helpers rushed off to a contest at the General Motors proving ground in Michigan. Their car put out a trace of pure nitrogen along with water vapor. Overall, the car cleaned up air pollution instead of causing it—there were fewer hydrocarbons coming out the exhaust than existed in an equal amount of surrounding air!

An Environmental Protection Agency representative offered him a grant, as long as Billings had a place to do the research. But Brigham Young officials said, "No, you've graduated and are not on the faculty, so we can't accommodate you." So Billings decided to form a not-for-profit research institute. He applied for tax-exempt status, but the Internal Revenue Service also turned him down. Undeterred, he formed Billings Corporation, a private research group. After a grant provided money for the start-up costs, Billings Corporation had a steady source of income—it made one of the first microcomputers in the country, back in the days when Apple was just becoming established. The company also became co-owner of the patent for the double-sided floppy disk. The cash flowed from the computer business into hydrogen research.

As in the case of Francisco Pacheco, the oil embargo of the 1970s helped propel Billings's work forward. The Postal Service gave Billings Corporation a mail truck, which was converted to hydrogen and used to deliver mail in Independence, Missouri, for a year. The only thing that stopped its use was the cost of the fuel—25 percent higher than that of gasoline. (Billings hopes that he can try again with the Postal Service, since recent technological advances will allow the trucks to run cheaply.)

In 1977, the Billings team converted a bus to hydrogen for the California city of Riverside, a project funded by the state's Department of Transportation. The project got off to a difficult start. The team did not get as much money as they had requested. Of the money that was appropriated, the city took some money to cover its administrative costs, and some of the money was used to buy the bus. The Billings team was left with $61,600 out of the original $125,000 appropriation.

The problems continued when the bus stalled because of silica in the carburetor. Since no silica was used in the fuel system, Billings concluded that the bus had been sabotaged, and had to send a technician to California to protect and maintain the bus. By that time, people involved in the project were discouraged, especially since the hydrogen itself was blamed for the problems.

Billings learned from this experience. First, not everyone shared his enthusiasm for hydrogen technology. But more importantly, he learned that making the fuel tank out of aluminum cut down on the weight, which made the technology much more cost efficient.

Billings went on to create a Hydrogen Cadillac Seville, which rolled to fame in President Jimmy Carter's 1977 inaugural parade, and the Hydrogen Homestead, a home he built for his family. It featured a kitchen range, barbecue, fireplace, and lawn mower that ran on hydrogen. Solar panels on the roof helped make electricity for his efficient hydrogen-generation system, and the home was heated with a hydrogen heat pump.

Billings firmly believed in the late 1970s that the United States government would give his research the support it needed to really get someplace. But the government did not. What happened?

What happened was the end of the oil embargo. Lines at service stations dwindled, and so did government funding of alternative fuels. "People went to sleep; they forgot about alternative fuel and pretty soon it was back to business as usual," Billings recalls. His company had gone public, and shareholders did not share his devotion to the hydrogen dream. Instead, they wanted to divide the profits. So in the late 1980s, Billings sold his interest in the company.

The LaserCel: Moving Ahead

Billings and his wife, Tonja, decided to put the money from the stock sale into developing a more efficient hydrogen fuel cell, a cell that would be lightweight and inexpensive enough for a car.

A fuel cell is basically a box in which hydrogen and oxygen combine to form water vapor. This process frees energy in the form of electricity. Part of the electricity, in turn, is used to keep the process going. When run in reverse, it can use electricity and water to create hydrogen, which can be used to recharge the hydrogen fuel tank.

Fuel cells were developed in the 1960s for the space program, which now uses them to provide electric power aboard the shuttle. However, these fuel cells are expensive and bulky.

Billings decided that fuel cells could be made lighter and cheaper. So in the mid-1980s, he assembled a team of researchers at the American Academy of Sciences in Independence, Missouri, and started working on his car-sized fuel cell. After many tries, his group finally came up with a method that used a high-powered laser to make a very compact cell.

Billings couldn't get any money from the federal government, but did receive some funding from the Pennsylvania Department of Energy for a next-generation fuel cell. In the early 1990s, the Billings group was ready to take the LaserCel 1, the world's first fuel-cell automobile, to trade shows. The hydrogen is stored in a 300-pound metal hydride tank in the back of the vehicle. If the car has a hydrogen-powered internal-combustion engine, such a tank would allow a range of 150 miles. But with a fuel-cell system powering an electric motor, Billings says that the same tank can allow a

450-mile range. "We solved our weight problem with the hydride, by using the hydrogen three times as efficiently. We also solved our cost problem."

The LaserCel car is inexpensive to operate because it doesn't waste energy like an internal-combustion engine does. More of the energy generated by a standard engine goes out the tailpipe than into moving the car. But most of the energy generated by the LaserCel goes into powering the vehicle.

The LaserCel provides heating and air conditioning effortlessly—passenger heat comes from running exhaust from the fuel cell through a heat exchanger, and cold is produced when hydrogen is taken out of the hydride. There is an accelerator battery to add power when speeding up, a battery that is recharged while the vehicle cruises. And the little LaserCel is very driveable. One newspaper reporter who tested it said, "Gee, it's nice to drive an alternative-energy car that goes fast enough to get a speeding ticket."

Billings estimates that the hydrogen system would cost about $4,500 a car if manufactured as part of a 10,000-unit run (car body extra). When mass-produced in numbers such as Detroit deals in, he believes the price of the car would be competitive—about the price of a midrange sports car.

Why hasn't this planet-friendly technology been mass-produced yet?

"Unfortunately, the power of the large vested interests in oil seems to have a stranglehold on the people making legislative decisions," says Billings. He says that he tried paying for a Washington lobbying office to try to get Congress to support hydrogen research. Hydrogen legislation was drafted and submitted, only to be lost somewhere in committee. No politician wanted to openly vote against such legislation, Billings speculates, so it never got to the voting stage.

The American Academy of Sciences is now the International Academy of Science, and is supported by small contributions from people all over the world. Billings and the academy are involved in a wide-ranging venture called Project Hydrogen. The academy has acquired millions of square feet of underground caverns—mined-out limestone—to finish into dormitories, classrooms, computer and fuel-cell laboratories, and metal-working shops, as well as offices. The academy offers graduate degrees in research, based on a "people learn by doing" philosophy. The student body is not large, under a hundred.

Project Hydrogen is exploring a number of novel technologies. For example, Billings says that blue-green algae, tiny plants that have been dismissed as mere pond scum, can do the same work as electricity does in electrolysis—separating water into oxygen and hydrogen. And after the algae finish doing that, they can be dried and eaten. Bottles of powdered blue-green algae have been on shelves of health food stores for years, since they are a concentrated source of protein and minerals. "You can have your energy and eat it too," Billings says.

Roger Billings's mission to develop hydrogen power continues. He has cheerfully endured his share of failed experiments; an inventor has to keep trying until something clicks. But the stakes are too high for him to not go on.

In the next chapter, we will meet inventors who have can turn heat that nobody wants into electricity.

10
Turning Waste Heat Into Electricity

All the major breakthroughs come from small guys in back rooms somewhere doing the impossible, because the big guys know it's impossible and they've got this rulebook that says what will work and what won't work.

—Les Adam,
Manufacturer

I think low-temperature phase change is going to be the solar power of the future.

—Peter Lindemann,
Energy consultant

The key to a self-running engine could be hiding behind your refrigerator. It's called heat technology.

In the nineteenth century, Nikola Tesla, the father of new energy we met in Chapter 2, started looking for the best way to solve humankind's energy problems. Just after the turn of the century, he wrote a paper in which he examined various methods of drawing energy from the environment. He concluded that the solution was to harness the energy that exists in the heat of the sun-warmed air surrounding us. Tesla saw the atmosphere as a vast source of energy, and he spent about twenty years trying to develop an engine—he called it a self-acting engine—that would run on this clean, plentiful energy source.

In this chapter, we will first learn how this technology works, and why it is so promising. We will then meet some of the inventors who are trying to bring this technology to market.

GETTING ENERGY FROM HEAT DIFFERENCES

New-energy historian and consultant Peter Lindemann of New Mexico has spent about twenty years looking into energy alternatives, and has found Tesla's ideas for a self-acting engine to be the most promising: "Tesla figured that if he could devise a way of creating a cold spot that he could dump heat into all the time, he could develop a way of extracting energy from that [difference in] heat." Taking energy out of heat by converting it into mechanical or electrical energy is everyday science—the steam engine is a good example. But in everyday science, these engines require fuel.

Lindemann believes that most Tesla researchers have not really understood what Tesla meant by a self-acting engine. Many assume that Tesla meant magnetism or gravity acting alone could be used to drive machinery. Tesla said these approaches were possible, but not really likely to work. On the other hand, Lindemann says, Tesla assumed it was wrong to immediately decide that the laws of thermodynamics apply in all cases. These laws put limits on the amount of power we can get from a machine, compared with the known amount of power we put into it. The assumption in the second law of thermodynamics is that if we want the temperature in a house to be either warmer or cooler than the temperature outside, we have to use up energy to make it so. Tesla did not accept this limitation as being universally true. Tesla thought that if living organisms could take energy from the surrounding environment, why couldn't machines do the same?

Heat Technology: A Refrigerator Running in Reverse

Lindemann demonstrates Tesla's concepts by opening the refrigerator in his kitchen: "This machine is the first cousin to Tesla's self-acting engine. We have to put energy into this machine to create a cold spot."

A refrigerator uses energy very efficiently—for every watt of electric energy it consumes, it moves three times as much heat energy into the surrounding air. This efficiency represents a potential energy surplus. Tesla knew that such a machine is reversible—that it would be possible to start with a cold spot and derive energy from it. Such energy would be free in the sense that a consumer would not pay a utility company for it, since the energy would come from the sun and be stored in the atmosphere as heat. Tesla designed his engine to use that surplus energy to do mechanical work. However, he was unable to build a working model because other technologies, such as refrigeration apparatus, were not developed enough in Tesla's day to support this self-acting engine.

Modern new-energy engineers have continued where Tesla left off by building a machine that uses refrigeration to produce mechanical energy. Today, it is called Low-Temperature Phase Change technology (LTPC), because it works with fluids that change from a liquid to a gas at a low temperature. It could be built as a unit that would both power and air-condition your home. This would make it both environmentally friendly and decentralized—no more hook-up to your local utility company, unless you want to sell your excess electricity or buy power when you splurge on electricity.

Lindemann says a power plant that uses heat as its power source could help to solve the global warming problem—the rising temperature of the earth's atmosphere caused by the burning of fossil fuels. "We could turn [this heat] into a major resource . . . and use it to rebuild the environment."

Heat Technology and the Marketplace

Despite some problems, LTPC technology is being developed along several different lines. A potential setback occurred when fluorocarbons, the gases used in refrigeration units, were judged to be harmful to the earth's protective ozone layer, and were banned by several countries. However, other refrigerants are being developed.

LTPC technology research is being done around the world. In Germany, Bernard Schaefer, Ph.D., has been granted a patent for a machine that produces—rather than consumes—electricity while it refrigerates. Schaefer has experimented with fluids that boil at temperatures below the temperature of the surrounding air.

Ocean Thermal Energy Conversion (OTEC) is another example of heat technology in use. This technology uses the difference in temperature between the warmer water at the ocean's surface and the cooler water below—a difference of about twenty degrees. A test model, on a large barge off of Hawaii, produced 50,000 watts.

Lindemann says that a number of LTPC technology researchers are very close to having commercial products on the market. Why is this technology so advanced, compared to some of the other new-energy technologies we've seen in this book?

One reason LTPC technology is so close to a marketplace breakthrough is the low maintenance such devices require. How often do you have to service a refrigerator? Lindemann points out that his refrigerator is the most reliable appliance in his house—he plugged it in, put it to work, and forgot about it. There aren't a lot of moving parts to break or wear out. This low-maintenance aspect of LTPC technology would be an asset in a power plant, as utilities now pay a lot of money for maintenance.

Another one of LTPC technology's advantages is its efficiency. It is potentially 400 times more efficient than the solar photovoltaic systems—solar panels that turn sunlight into electricity—that are on the market. LTPC technology is also more efficient than nuclear power. Lindemann says, "You can get more power from a system like this than you could ever get out of a nuclear reactor. These things can be scaled up to megawatt [power levels]."

If a LTPC system is built to use direct sunlight, its efficiency rises dramatically. However, Lindemann is most impressed with systems that don't need direct sun, because they will run at night, even in colder, cloudier climates.

GEORGE WISEMAN AND THE LTPC HEAT PUMP

One of the inventors leading the rush to the marketplace lives far away from the big consumer markets. George Wiseman, a cheerful-but-preoccupied man, lives in British Columbia on a farm in the Rocky Mountains, when he isn't travelling around the world as an energy consultant. There, surrounded by forests, the self-taught electronics expert writes how-to books for new-energy enthusiasts and does experiments in his rough-hewn workshop.

Wiseman has put LTPC technology to work in the form of a heat pump, which is a machine that takes heat out of a substance that is at a slightly higher temperature than its surroundings. For example, dumping hot effluent from a factory into a river raises the river's temperature, producing heat that can be captured by a heat pump. Heat pumps themselves are nothing new. They are currently used by some affluent, environmentally conscious homeowners to take heat out of the earth for home-heating purposes.

A heat pump uses a fluid that changes phase from a liquid to a gas at -40 degrees Fahrenheit or colder. In contrast, a standard steam engine runs on a fluid that changes from a liquid to a gas— steam—at 212 degrees Fahrenheit. So instead of using boiling water, the heat pump uses low-boiling-point fluids.

George Wiseman works on his inventions, including a new-energy heat pump, in rural British Columbia.

Phase change is a relatively simple concept. A liquid can either freeze into a solid or evaporate into a gas. Solid, liquid, and gas are known as the *phases* of a substance, and when a substance changes phase it either contracts or expands. The expansion phase creates pressure that can drive an engine. It is more efficient to use a low-boiling-point fluid than it is to use a high-boiling-point fluid such as water because the former can take heat from almost any environment, no matter how cold.

But as environmentally sound as a heat pump is, it still wastes energy—heat is lost through the pump's condenser, the part that cools the gas back into a liquid. Tesla designed a heat pump that had no condenser, and thus eliminated the part that throws away the heat. Building on Tesla's design, Wiseman's LTPC system recycles unused heat to make the system more efficient. It can take heat from air, earth, or water and convert it into electricity.

Wiseman's system has two interesting features:

• A cheap, efficient collector of heat from the sun—a radar-like dish covered with reflecting film.

• A design that relies on inexpensive materials, so that a backyard mechanic can build a one- to five-kilowatt home power system, including batteries, for under $1,000—a tenth of what it would cost to install a standard solar setup.

The Wiseman heat pump sounds good for a homestead, but what could city dwellers do with it? Wiseman envisions a number of uses, since the pump can tap into large sources of power wherever heat is stored, such as oceans and lakes—including manmade lakes. This means, for example, that Wiseman's pumps can be used to cool a city's reservoir, lessening the loss of water through evaporation, at the same time it makes electricity to run the city's air conditioners.

Wiseman also sees his technology being used in industry. A factory could use the waste heat it generates during various manufacturing processes and turn it into electricity. This would lower the factory's costs, and thus make it more profitable. Such cogeneration of electricity means that the factory could also sell power back to the electric company, further improving profits.

HAROLD ASPDEN AND THE STRACHAN-ASPDEN DEVICE

British physicist Harold Aspden, Ph.D., is as busy in retirement as he has ever been, and his new-energy career keeps gathering momentum. The theories he had spent decades developing in his spare time are the subjects of enthusiastic articles in new-energy journals. Aspden's credibility is enhanced by his earlier careers: nineteen years as IBM's director of European patent operations and, after early retirement, nine years as an IBM-sponsored visiting staff member of the University of Southampton's Department of Electrical Engineering.

Though he is now retired from the university, Aspden still has his hands full. His research enterprise, Thermodynamics Limited, won a United Kingdom government grant for tests on a new kind of electric motor. He also writes on new-energy issues, and send items to a new-energy newsletter to update all his associates across the globe.

It is not surprising, then, to learn that Aspden's ideas carry weight in the new-energy world. And the idea that is getting the most attention relies on the power of magnets, a power we first learned about in Part II. Aspden is known for his innovative magnet-motor research, and also for furthering the idea that the power of magnets can be used to create heat technology for noiseless

Dr. Harold Aspden of England is a very active new-energy researcher. He developed, along with John Scott Strachan of Scotland, a device that uses heat to generate electricity.

refrigerators and air conditioners that don't use ozone-depleting gases.

Magnets and Heat Technology

Aspden, who lives in England, had been exchanging ideas long-distance with engineer John Scott Strachan, who was working for the American corporation Pennwalt in Scotland. It had been a very fruitful exchange for both men, but now they needed to get together and talk. Instead of one driving 400 miles to see the other, however, the Planetary Association for Clean Energy in Ottawa, Canada, solved the problem by inviting them both to speak at a 1988 conference.

A long delay in an airport is rarely pleasurable, but Aspden and Strachan were on an intellectual high on their way back to Britian after the conference. As they sat in the busy terminal, they both saw a clear challenge: where do we find super-efficient alternate energy sources? Like Wiseman and others, they wanted to make electricity from a very slight difference in temperature, using a method of refrigeration that would replace all the ozone-damaging fluorocarbons that are normally used.

Aspden and Strachan decided that the answer was to develop a heat technology invention they had been working on. At Pennwalt, Strachan had been sandwiching a plastic-like material between metallic films. He wanted to see if the resulting assembly, which produced an electric charge when pressed, could be used in a medical device he was working on.

Strachan found that the assembly was too heat-sensitive for that use, but he and Aspden thought that they could add magnetism and use the magnetized assembly to create a unique form of electricity-generating thermocouple. In a thermocouple, two different kinds of metal are joined, or coupled. The fact that the metals differ creates a voltage, or electric pressure, at the joints, which are called junctions. The amount of voltage varies with temperature of the junction, so thermocouples are generally used in science and industry as very sophisticated thermometers.

The voltage at the junction of a thermocouple is too small to be used as an efficient source of electricity. However, Aspden and Strachan found that they could combine Strachan's plastic-and-metal sandwiches and Aspden's knowledge of magnetism to create a device that used the difference in temperature to generate useable amounts of electricity from fairly small inputs—even from a melting piece of ice.

The airport meeting gave them time to talk about how to test their idea and to map out a development strategy. The invention had to be protected by a patent before it could be publicized. Pennwalt would have to formally release to Strachan some rights to the invention. As the patent expert, Aspden wrote two patent applications, each covering a different aspect of the device. Meanwhile, Strachan left Pennwalt to build test devices.

The duo formed Strachan-Aspden Limited to hold the patent rights to their apparatus. However, the United States patent did not come easily. At first, the patent examiner flatly stated that it was impossible for the Strachan-Aspden device to make electricity. But, finally, he accepted the pair's evidence.

What could the Strachan-Aspden device be used for? It could be used by industry to turn waste heat into electricity. But it could also be used in homier settings. A family with a greenhouse could use the difference in temperature between the air inside the greenhouse and the air outside to make electricity to run household appliances. Thus, the device would allow individuals to generate at least some of their own electricity.

Aspden and Strachan's device could also pave the way for the development of practical superconductivity. Superconductivity happens when a current-carrying material, such as a wire, suddenly loses resistance—the force that keeps the electric current from passing through it. This allows a nearly perpetual passage of current through the material.

Superconductivity normally takes place only at very low temperatures, and is normally used to create powerful electromagnetic fields. However, the Strachan-Aspden device could allow superconductivity to take place at normal room temperatures. This, in turn, would allow superconductivity to be used in creating a practical source of electric power.

Going Separate Ways

Although heat technology research excited Scott Strachan, he eventually turned back to his original specialty of designing optical measurement hardware. It wasn't as exciting, but it was more predictable—each of the three working heat technology models Strachan had built had deteriorated with use over a period of months. Why? Aspden's guess is that heat and vibration destroys the device's electrical storage capacity, interfering with its operation. Better materials are needed, and both Strachan and Aspden expect advances in this field to move their technology forward. They agree that the invention is sound in principle and that it works, but that it needs a well-equipped corporate laboratory to develop the materials to the point where the device could be produced commercially.

Because of the difficulties of long-distance collaboration, Aspden and Strachan decided to form separate companies. In Scotland, Strachan-Allen Limited concentrates on Strachan's optical research, while in England, Thermodynamics Limited does the heat technology work. Aspden bought patent rights to the Strachan-Aspden device, and later added other patents. But the two inventors stay in touch. For example, Strachan is involved with a branch of an English optical research company that has business ties to Aspden.

While Aspden works on another device, a magnetic motor designed to generate electricity, the Strachan-Aspden device sits in limbo. The lone researchers Aspden and Strachan were able to take it to the demonstration stage, but it now needs a corporate champion to turn it into a commercial product. Aspden says that there's a twofold problem in finding the needed support:

• The scientific community does not believe the device will work.

• Nonscientists see this high-tech development as being too complicated to understand.

He adds, "The new-energy dream of power generation from environmental heat is already with us, but we have somehow ignored Nature's message."

In the next chapter, we will learn how new-energy researchers are finding gentler ways to use a traditional method of power generation—hydropower.

11
Low-Impact Water Power— A New Twist on an Old Technology

What the public doesn't realize is that bias [of those making energy choices] and conflict of interest are keeping out viable sustainable energy options that create jobs.

—Nova Energy Ltd. press release

Most of our real innovators are outside the system.

—Martin Burger, managing director of Nova

For years, hydropower has generally meant massive dams and environmental dislocation. Even the few small-scale hydropower projects that have been built are bad news for fish trying to swim upstream, while both large and small dams degrade the quality of the water downstream. And hydropower doesn't have to involve a standard dam to be destructive: the Bay of Fundy tidal power project on Canada's southeast coast has devastated the home of countless sea creatures.

But now, there are water-based technologies that work with nature and not against it. Viktor Schauberger, the water-energy pioneer we met in Chapter 3, would probably applaud. One creates its energy effects by spiralling either water or air within Schauberger-inspired generators. William Baumgartiner builds prototype machines that show there is an alternative to today's dirty-energy technologies based on either pollution-spewing fossil fuel or river-destroying dams.

Another revolutionary alternative can be built as individual modules that can be placed in a current of water anywhere, from a slow-moving river to a tidal basin. Strung together in connected units, these modules can add up to megaproject output without megaproject impact—or megaproject cost. One of this technology's advocates describes it as being "in harmony with humanity's existence on the planet . . . refined in comparison with the rough and crude engineering we have today." If Schauberger were alive today, he would understand Martin Burger's passion for the Davis Hydro Turbine—and Burger's battles against an outdated megaproject mentality.

MARTIN BURGER: LIGHT FROM THE WATER

As a Cree child living along the Mackenzie River in Canada's Northwest Territories, Burger learned from shamans—medicine men—and others the value of being in harmony with nature. He will never forget the traumatic years of 1963 and 1964, after the provincial government built the first dam on the Peace River, whose waters eventually reach the Mackenzie. Water levels were low in the big river. But the amount of water was not all that had changed; a subtle quality had also ebbed.

"The elders were in despair. They talked about how 'the light had changed with the river.' That river was the centerpiece of the community; it was the highway, the life-giving force," recalls Burger. "Now, the river's 'light' doesn't mean anything to a material culture like ours. But it meant everything to that culture.

"There is a dimension to the vitality of a river that we don't appreciate when we build dams. . . . When they bemoaned the fact that the river was 'dark,' they meant that a part of their landscape died and they were all wounded. We may not understand that in our Western culture for the next two or three hundred years."

After his father moved to a mining town in 1965, the teenaged Burger plunged into the world of machinery, and supported his university schooling by working in a mine. He trained as a civil engineer and worked with Dow Chemical and other multinational corporations before going back to mining in the Territories.

It was during 1988 and 1989 that Burger's quest for a new-energy technology really began, after a change in the business climate suddenly jeopardized the silver mining operation he was running. He was $3 million into the $8 million Arctic Circle operation when the Canadian federal government changed the tax rules, which caused sources of capital to flee from the project. Burger was left stranded.

How could he cut costs dramatically in order to salvage the situation? Storage tanks for a million gallons of diesel fuel stared Burger in the face, reminding him of the $4.25 a gallon he paid to fly fuel into remote Great Bear Lake for the generators. This added up to more than $4 million a year to keep the lights on and to pump the water needed for the milling operation.

Going With the Flow

Something caught Burger's attention—the motion of water flowing at a speed of about seven knots near the pumphouse. He knew that was where the answer must somehow be found: "I knew there had to be a way, even if I had to fabricate an old paddlewheel. If I could turn a shaft, I could turn a gearbox and run a generator."

He asked around for advice, and Canada's National Research Council (NRC) put him in touch with Barry Davis, a world-class engineer and designer. Besides designing planes for the Canadair and Bombardier aircraft companies, Davis had also developed an innovative ship design for the DeHavilland Aircraft Corporation of Canada. In 1969, he put a tiny, V-shaped hydrofoil on the bottom of a 225-ton Canadian Navy destroyer. This winged hydrofoil allowed the ship to cruise at sixty-five knots or more twelve feet out of the water—a lot faster than a destroyer's normal twenty-knot speed.

Although the ship project didn't go into production, the performance led to Davis's idea of hooking the wing to a shaft and turning it in a circle to generate torque, or turning power. The torque on a shaft would drive a gearbox, which would drive a generator, which would produce electricity—all without a dam.

The genius of the simple design is that the blades rotate faster than the water moves over them, whatever its direction—water flows over the blades like air flows over an airplane wing, creating lift. Another of the turbine's strong points is that fish can swim safely through the slow-moving, rounded blades. Burger was impressed.

Davis had formed a company, Nova, in the mid-1970s to develop the turbine because he saw a need for a gentler form of hydropower. The NRC was impressed, enough to give Davis's company the money to build and test three prototypes between 1978 and 1988, at which time the newly elected Conservative government cut the council's budget. In 1989, Davis joined forces with Burger, whose

The Davis Hydro Turbine, built by Nova Energy Ltd. of Canada, uses running water to create power— but unlike conventional hydropower equipment, it does so without creating environmental problems.

mining operation eventually went dormant, and the company was reborn as Nova Energy.

Meeting Resistance

The effort that went into creating the Davis Hydro Turbine was almost nothing compared with the effort that has been required to sell it, at least within Canada. Burger, working from Vancouver, British Columbia, as managing director of Nova, has experienced a series of frustrations in attempting to get the local utility, B.C. Hydro, to give the environmentally benign turbines a serious try. He has also tried to interest B.C. Hydro in the job-creation potential if the province were to export tidal power machinery. The company has rejected his proposals.

If the Davis turbine hasn't drawn North American sales, it has drawn praise. The NRC says, "The Davis Hydro Turbine is sufficiently well proven and it's time to commercialize [its] development." And while negotiations with the United States Army Corps of Engineers fell through, the corps says that the turbine is "technically sound."

Is the technology too simple? Do high-salaried engineers fear for their jobs if it becomes obvious that complicated engineering projects are unnecessary? Burger says that the regional utility company's engineers would never admit to this publicly. "But we have a paper trail showing that they've deliberately provided misleading technical input into policy formation" by British Columbia's provincial government, which has since been advised by its own experts to act on this opportunity.

Hydropower has always been an engineering-intensive energy source, and has generally created widespread ecological disruption. River dams create lakes that cover thousands of acres and often lead to unforeseen problems, such as erosion and the loss of plant and animal life. The local human population is also affected. In tropical areas, there can be serious public health problems. Sometimes, the power produced by the dam is sold cheaply to large industrial users, with little or no benefit for individuals living near the dam. In other cases, power is sold so cheaply to local customers that demand outruns supply, leading to the need for even more power plants.

The Davis turbine could make damaging megaprojects such as dams obsolete. It holds the promise of powering a three-bedroom home with a fuel-less generator small enough to fit in the back of a pickup truck. Or, according to Burger, the low-impact turbines could eventually replace nuclear power plants on the Eastern seaboard by being strung together to create a megawatt power plant in the currents of the Gulf Stream. Burger explains that we could have had this kind of low-impact energy generator about ninety years ago, when such a device was first invented, but that the decision was made earlier in the century to go with the now-standard high-head dam, in which water is held behind a high wall and allowed to fall over turbines.

Nova says that wherever water moves at a speed of from two to twelve knots, the Davis Hydro Turbines can be plunked into the water and put to work. The Davis turbines need only water speed, not water height, in contrast with projects such as the dam in southeastern Canada's Bay of Fundy. That out-of-date technology alternately traps and releases water with a low-height dam, a wall that blocks the natural flow of silt and wrecks ecosystems.

Despite the lack of interest by the big power companies, Nova keeps on trying. Burger and Davis would like to sell small units for home and business use. But because Nova needs money to finance the turbine's final premanufacturing stage, the company is forced to look for large contracts first.

Will Nova ever be able to carve out a market in Canada? Not only is Canada a bastion for conventional hydropower, but the country also puts its money into fossil-fuel and nuclear power, and inventors of other energy sources get the runaround from officials. One newspaper columnist writes, "Here, the R and D is focused on conventional power generation, not the future of emerging energy technologies."

Perhaps B.C. Hydro—which a frustrated Martin Burger sometimes calls "Hydrosaurus Rex"—will eventually decide that if it can't beat the energy revolutionaries, it might as well join them. After all, some new-energy visionaries are willing to put their unpatented knowledge on the Internet in order to turn the technological tide away from destructive megaprojects. For example, Burger and Davis have decided to put a small-scale version of the Davis turbine on the Internet to help the planet.

What led these men to make such a generous decision? For Burger, the goal is getting clean, affordable power technologies to people who struggle to survive. For example, he has been contacted by the Dene, an aboriginal group in the Northwest Territories, many of whom are trappers based in small villages along the rivers. Burger dreams of seeing the Dene manufacturing the Davis Hydro Turbine and reaping the economic benefits.

A Perfect Match: Low-Impact Turbines for Poor Nations

Low-impact turbines are a good energy choice for developing countries that do not have megaproject financing. Some countries, such as Nepal and China, are already committed to a program of small-scale hydropower plants. Such units, scattered throughout the countryside, make the lives of rural residents easier by taking over such drudgery as hulling grain and pumping water, and by providing power for the small businesses that often develop near the plants. This, in turn, keeps people from migrating to the crowded cities.

Burger says that the Davis turbines are a good choice for developing countries because a number of units can be linked together into one large power plant, and thus provide power to a larger area. However, a country would not have to purchase all the turbines it wanted all at once. It could start with as many turbines as it could afford, and add one at a time. If selling large projects goes well, Burger wants to give away smaller units to villages.

The buying plan proposed by Nova sounds as efficient as the generators themselves. Burger gives an analogy: take someone working for minimum wage and give that person a chance to buy a $400,000 home. Instead of being faced with buying it in one crushing purchase, however, the worker can get a foot in the door by buying the house module after module: first the mailbox, then the doorsteps, then the front hallway, and so on. Each addition would be bought whenever the buyer had saved some more money.

In the same fashion, a developing country could start right away to generate electrical power by gradually building up its generating capacity. Nova plans to sell modular units—single turbines that generate power and, therefore, revenue as soon as they are installed. Additional units can be bought when more money is available. A country could conceivably have a megawatt-generating platform in about the time it would have taken to build a conventional megawatt power plant, but without paying the high interest charges on the money it would have had to borrow to build a conventional plant.

The Philippines may take the clean-energy route with Nova's proposed generating station across the Hinatuan Passage between two of the country's southernmost islands. Under the proposed financing plan, the country would risk only $150 million in corporate loans—small change, compared with other energy options—for a single tidal power module at any one time. "As soon as you have built the first module, the water starts turning these rotors and it generates a return on the investment," Burger says. The first module's output helps pay for the second module, which in turn helps pay for the third, and so on. The country would end up with a billion-dollar generating station—without a billion-dollar debt. Nova is also looking at a tidal power project for the Indian government in the Gulf of Cambay, near the Pakistan border.

WILLIAM BAUMGARTINER: THE IMPLOSION GENERATOR RETURNS

If the Davis turbine represents a gentler version of twentieth-century energy technology, an implosion generator represents a leap into twenty-first-century technology. Viktor Schauberger started exploring implosion technology earlier in this century, creating generators that worked on radically different principles than do standard power plants. William Baumgartiner of Albuquerque, New Mexico, is carrying on the Schauberger legacy.

Baumgartiner was born in Germany, and was schooled there and in Switzerland, where he hiked and skied in the Alps. Like Schauberger, he spent hours beside mountain streams and learned by observing natural systems. But school bored him, especially when teachers deflected his questions by quoting authorities from the past. He continued to question dogma in science, but did go for a higher education—a mechanical engineering degree from Technisches Institute in Zurich, Switzerland.

In 1954, he moved to Canada with a dream of living off the land in a trapper's hut out in the bush. Reality, however, forced him to work in a sawmill and study English in his spare time. He then landed a job as an electrician on power dam projects.

Baumgartiner came upon books such as *Living Water*, a biography of Schauberger, and would eventually see the importance of letting a river run naturally in unfettered spiralling movements. Like Burger and Davis, he would end up looking for workable alternatives to conventional hydropower.

Learning From Past Masters

Baumgartiner's odyssey as a maverick engineer began when an acquaintance from Vancouver introduced him to Nikola Tesla's patents (see Chapter 2). At that time, few people knew about the electrical pioneer—the wealth of Tesla material now in bookstores was not yet available. A new job, as supervisor of a natural gas pumping station, gave Baumgartiner the time to repeat some of Tesla's experiments.

Baumgartiner then moved to the interior of British Columbia, where he supervised an automated pumping station from home. By that time, he had discovered the work of inventor John Searl of England, and used the workshop his employer supplied to build two Searl energy devices.

His friend from Vancouver, the one who had introduced him to Tesla's work, wanted to produce more Tesla memorabilia. He and Baumgartiner agreed that money would be needed for research, so they decided that Baumgartiner would write and sell information on Tesla technology by mail order. They advertised in *Popular Mecha-*

nics, and, to their surprise, up to forty letters a day addressed to the Tesla Research Centre showed up. This led to another discovery for Baumgartiner when one person sent him information about Walter Russell (see Chapter 3), with a note that Russell had occasionally met with Tesla in New York. Baumgartiner was impressed with Russell's knowledge of the invisible mechanics of the universe.

The more Baumgartiner learned about new-energy researchers of the past, the more he wanted to know. His many projects included publishing a magazine, teaching, and building nearly every unorthodox energy device ever invented. But time and again, he came back to Viktor Schauberger. As he read and worked, he began to grasp the principles behind Schauberger's inventions.

Schauberger had studied how river water is swirled by the rotation of the earth until a subtle electrical charge is concentrated towards the center of the river. Electrical charges build up when slightly different materials come into contact, and a river consists of layers of water with differing qualities. These layers twist inwardly in a spiralling movement known as a vortex. (See "Spirals of Energy" on page 13.) The twisting motion is like braiding relatively weak strings together to form a relatively strong rope; when this motion occurs in a river, the energy strengthens. Schauberger saw that the water acquired more energy when the inward-spiralling force was greater than the outward-spiralling force. He then used this principle to create his suction turbine, in which specially twisted pipes were spun around a central shaft until either the air or water within sucked itself through the pipes.

"I had only seen pictures [of the turbine] and only guessed at what equipment was inside," Baumgartiner says. "Then more and more, from what I knew of space geometry, [I] formulated in my mind how it should be."

Capturing Tornado Power

In the late 1970s, Baumgartiner began to develop what he called Twister Pipes—his version of Schauberger's equipment. The pipes, made of fiber glass and copper, are difficult to fabricate because of the odd shape needed—they not only spiral, but gradually become smaller and smaller on the inside surface. To make it even more difficult, a cross section of the pipe is not round, but instead resembles the outer edge of a human ear.

William Baumgartiner of New Mexico is building an implosion generator on the principles discovered by Austrian Viktor Schauberger earlier in the century.

Once he perfected his Twister Pipes, Baumgartiner started arranging them inside of a turbine. The theory is that the pipes create a three-dimensional spiral of tremendous energy—in effect, a tornado. When the air or water moves through the pipes, its inward-spiralling movement pulls the air or water forward at a tremendous speed. This force can turn the shaft of a machine, and thus do useful work.

The first tests of Baumgartiner's odd-looking water turbine showed overunity—more power going out than coming in. He believes this energy comes from space, as explained in Chapter 4. He also believes that current technology works against such energy: "With our machinery today we dissipate this force; it never accumulates or flows steadily."

How can we get this energy to work for us—to flow continuously instead of being scattered and destroyed? Vortexian movement—the three-dimensional spirals captured in Baumgartiner's turbine—seems to be the secret. Both Schauberger and Russell had pointed out that vortexian movement is often seen in natural systems, from the flow of blood to the flow of rivers. Baumgartiner describes the vortex as "nature's tool with which it creates anything it chooses." He says that the inward-spiralling vortex catches the aether, the background sea of energy discussed in Chapter 4, and winds it faster and faster, tighter and tighter, until it creates an effect that human beings can detect. For us to use this energy, we must channel this force so that it not only flows steadily in from space, but is allowed to flow steadily out. We need to create a closed loop of energy.

He calls the process vortex mechanics, and says that it allows the building of "a living machine," a machine that can capture the life force of space. On the other hand, standard machinery is dead because no energy accumulates within it—"We burn fuels, they expand, and then it's over with." Such machines act in accordance with the laws of entropy, which says that matter and energy gradually fall apart and become increasingly disorganized. But living machines like Baumgartiner's act according to the laws of negentropy, which says that matter and energy can organize themselves. It is like a wheel on a hill; under conditions of entropy, the wheel will only roll downhill, while under conditions of negentropy, the wheel could roll uphill. Proper use of negentropy knowledge means that humankind can move past the destruction of matter and dissipation of energy into a future when energy will flow continuously, without harming the environment or causing the hardship of scarcity.

The time and money that Baumgartiner has spent on his turbine in recent years has cut into his income, but he persists. His work is gaining international attention—he has been called to Australia as an implosion-technology consultant. This technology is also being studied in Europe.

In the next chapter, we will meet other energy innovators who have tried to make the energy revolution a reality.

12
The World of Power Possibilities

Many important discoveries are made by experimenters who are not classified as scientists.

—Harold Fox,
Physicist

The main objective has been reached, namely to prove it is possible to use Free Energy.

—*Thesta-Distatica*, a film
about the Methernitha commune

Mysteries of water, harmony with nature, implosion technologies—we were first introduced to these themes when we met Viktor Schauberger in Chapter 3, and saw those themes continued in Chapter 11. Again, I find these themes in other new-energy technologies, such as those invented by the energy pioneers in this chapter. They reflect the variety of backgrounds found among today's innovators, but they also have something in common— they offer an alternative to traditional, explosion-based energy hardware.

Some of these inventors, such as the Austrian naturalist and the members of a Swiss commune, have kept their energy technologies under wraps in order to discourage strangers from visiting them. The American businessman, however, encourages test engineers to visit his manufacturing plant in Georgia. He wants science to find out why his hardware does what it does.

JOHANN GRANDER'S LIVING WATER

Is Johann Grander a Viktor Schauberger for the turn of a new century? There are similarities between the two unpretentious Austrians. Both men's energy-conversion devices focus on being in harmony with nature. And Schauberger had to sign away his patent rights before he was allowed to go home to Austria from Texas, while Grander could not even get a patent on his early energy invention.

Grander was born on April 24, 1930, in Jochberg, Austria, and only fourteen years later was out making a living. In 1962, he rented a gas station in Jochberg. But there was always something different about Grander. One of Grander's close associates, Austrian businessman Georg Huber, says, "Though he seemed to be a simple, hard-working man, he always was an *impressi[ve]* person. When he had the gas station, he was the political leader in his home village (emphasis in original)."

In the 1960s, Grander began to spend more time high in the mountains, communing with nature. He started thinking about pollution from gasoline vapors—what could be done about such health hazards?

He had more time to explore these ideas after 1974, when he went into business building cabins in the Alps, where the clean environment inspired his thinking. In 1989, by an incredible stroke of luck he acquired the *Kupferplatte* (Copper Plate) mine, the only privately operated copper and silver mine in Austria. Although mining operations had stopped in 1926, he and his associates turned it into a tourist site. There he found not only a source of income, but a tranquil haven, and began learning about mineral ores and cosmic radiation.

Grander's mechanical tinkering had begun years earlier, when his wife started suffering from back problems that doctors could not cure—she could not even work in her garden. Realizing that he had to rely on his own resources, he remembered that his father had said something about magnetism helping to cure ailments. So Grander built a magnet massage roller that helped both Grander and his wife heal inflammations. Although the device disappeared—they loaned it to someone else and never saw it again—Grander's wife felt good enough to work in the garden and gather berries in the mountains. Grander further explored magnetic forces.

Working With Nature

Again, like Schauberger, Grander found that water was a key to unlocking the door to nature's secrets. Grander called his discovery "living water."

Grander's water is taken from a deep spring in the mine and further energized by magnetic vibrations. What is different about it? Austrian journalist Hans Kronberger says that a water expert, Dr. Horst Felsch, tested Grander's revived water. The depth of a water spring is tested by checking to see if the water contains tritium. If no tritium—a radioactive substance that spread through the atmosphere after the first atom bombs were detonated—is found in the water, it means the water has been sheltered below the earth's surface since 1945. "All the water that has come to the surface [since 1945] is contaminated with tritium, worldwide," says Felsch. No tritium was found in Grander's water.

Another standard test involves growing bacteria on a water filter. Normally, the bacteria grow in a random pattern. But after Grander's water was tested, the bacteria were distributed symmetrically. Felsch says, "During the thirty years of my professional experience I have never seen anything like this." Felsch showed the filter to an electrical engineer, who said the pattern was the same as that found when iron filings are spread in water and a magnet is held close to them—they line up along lines of force leading to a north and a south pole. "This was the first scientific indication that there was a higher energy content in Grander's water," says Felsch.

Meanwhile, articles Kronberger wrote about Grander for *Sonnen Zeitung* (*Sun Newspaper*) magazine in Vienna brought a flood of positive testimonials from individuals, institutions, and businesses who said they were using Grander's water-revitalizing devices on water pipes. There were various claims: people said that health problems cleared up, a dairy reported more milk production, and industries were finding less scale in pipes. Laboratory testing shows that Grander's water is not changed chemically, but that its molecular structure is altered in such a way that harmful substances can be flushed out more easily.

Grander's living-water technology evolved along with his magnetic motor, which he developed over a period of decades. The motor requires neither a battery nor an electric outlet. In it, magnets are covered with special alloys, or mixtures of metals, and arranged in such a way that they continually excite one another to higher and higher frequencies. Grander says, "In this way, the natural magnetism of the magnets is intensified so much that . . . energy begins to flow." He says this is cosmic energy—"natural vital forces, the strongest of which you can touch without getting an electric shock."

Johann Grander, who claims that his machine creates a new form of electricity, is the favorite topic of many energy researchers in Europe. A European associate who visited Grander reports that the inventor plugged a hair dryer into the output of his generator and the dryer worked normally. Then Grander threw the hair dryer in a tank of water and it still ran, blowing a spray of water upward.

How did this unschooled man discover his revolutionary inventions? Grander is quoted as saying, "School is like a lawnmower, making everything equal." He believes that if he had not had to leave school early to help support his seven siblings, his thinking would have conformed to that of the science establishment, and that he would not have become an inventor.

Grander doesn't speak much about his development as an inventor, Huber says. "But we know he had visions. We know that when he was working with his motor and nothing worked, he was going up to sit up on the balcony in the sun, and then after ten minutes

Johann Grander of Austria (right) is the creator of what he calls "living water" technology, which works with natural forces to provide energy. Grander is shown with associates Dr. Horst Felsch (left) and Tat Chee Tam.

he heard an [inner] voice . . . telling him to go down and try it another way." Huber likens it to the story of Johannes Brahms. When asked how he created his famous compositions, Brahms said that when he was in harmony with nature, the music would come—not step by step, but all at once in one piece.

Grander Runs Into Trouble

Grander's joy at having come up with something that would help humanity soon vanished. After spending a lot of money to apply for a patent, he "was brutally shaken out of my dream" by an official notice he received in reply.

"'Inventions which are detrimental to products in existence may not be granted a patent' was the devastating decision of the patents office," Grander recalls. "My request for a patent had been turned down with . . . [the officials' statement that] 'this decision is adopted solely for the protection of the economy.'"

This barrier did not stop him. Grander had a supportive family as well as the feeling that he was being guided by God. Huber says that Grander found out that "there are other structures of power on this earth. But this didn't change his optimism; he was very sure that everything is according to God's will." Grander has continued through years of expensive research into what he called "natural and vital forces, healthy and inexhaustible sources from which our lives are forever surging."

Grander went into seclusion near Jochberg after a steady stream of visitors disturbed his family's privacy. Huber makes a point of saying that Grander has dismantled his motor, so that there is nothing for visitors to see. He does not travel much; his associates report that he refuses to get on a jet airplane because of the damage jet exhaust does to the upper atmosphere.

Grander is also unconcerned about money, despite the offers that have been made to him. Huber says that Grander turned down millions of marks from the German multinational corporation Siemens. He has also had no interest in lucrative offers from other countries, fearing that his revolutionary invention would either fall into the wrong hands—he says that military interests have been allowed to take over the most powerful technologies of the twentieth century—or be tossed into a filing cabinet and suppressed. Another of Grander's associates says that the inventor wants to release information on the motor when his inner guidance tells him that the time is right, so that all the people of the world benefit from it.

Despite his concerns, Grander is helping to cut down on the fossil fuel used in at least one country. In 1993, Grander's associate Tat Chee Tam, working from Hong Kong, convinced railroad officials in China to test a Grander invention called Eco-kat that uses concentrated magnetic energy. China Railroad Corporation, one of the largest rail networks in the world, uses more than 2 million tons of fuel a year—40 percent of all the diesel fuel consumed in China. In order to reduce both the fuel bill and the black diesel smoke, the railroad had spent ten years testing magnetic devices and other fuel-saving measures. Nothing impressed the railroad, according to the head of the engine research department, until Eco-kat. He says that Grander's liquid fuel treatment has reduced emissions, cut fuel usage, and increased horsepower.

A COMMUNE IN HARMONY WITH NATURE

Johann Grander is not the only inventor who has gone into seclusion. In 1960, Swiss technical wizard Paul Baumann fell in with a group of friends—people like him who wanted to work alongside others who cherished spiritual values, and who wanted to live together harmoniously. So they founded a no-tobacco, no-alcohol commune called Methernitha near Linden, Switzerland, a village surrounded by farms in a valley with dozens of farming villages. There, in that peaceful setting, Baumann and others from around the world started a number of cooperative cottage industries—such as building windows and making hardware for the building trades—as the economic basis for their idealistic community. Members used the savings from their enterprises to expand the commune.

From the beginning of Methernitha, electronics research and development has been a priority. The aim has been to develop an alternative-energy technology that works with, instead of against, natural forces. Solar heat collectors, waterwheels, and low-speed windmills were the first devices to attract the attention of the commune's researchers, who wanted a clean source of energy. Later, they concentrated on esoteric sources of energy—what we now call space energy.

Today, nestled among the brown mountain houses, the 200-member commune's businesses and land together are worth millions of Swiss francs. But the accumulation of wealth is not a goal in itself for the group. Financial independence simply gives them the freedom to make their own decisions. And among the rolling hills, the commune shelters one of the first self-propelled devices for harnessing space energy, which is discussed in Chapter 4. This breakthrough technology has been whirring away on a tabletop away from public view for years. The people of Methernitha want to keep it that way.

Methernitha Creates a Generator

In the 1980s, before the commune closed its doors to visitors, some European electrical engineers travelled like pilgrims to Methernitha. There, they inspected the Thesta-Distatika, the free-running, fuel-less electric generator that was built in several sizes, one of which was powerful enough to meet most of the electrical needs of a small home. The device was not connected to any battery or power line, or even to solar panels. A member of the community started the machine with a flick of his hands, by rotating two adjacent disks in opposite directions. After that, it ran by itself. Amazed, the engineers wrote reports and came back for repeated viewings.

According to one of the Swiss technical experts who saw the Thesta-Distatika—the machine he describes as a "tachyon converter," or space-energy converter—it is a masterpiece of craftsmanship and electrical engineering. He says the experience of inspecting it was breathtaking; he felt he was witnessing the phasing out of one era in technology, and the phasing in of another.

The Methernitha group explains in a public relations film why it has developed its "wondermachine." The commune says it wants to develop machines that "unlock sources of energy for the benefit of mankind without disturbing nature's ecological balance."

The film shows a hybrid energy-generating system—rotating electrostatic discs, and nonrotating magnets and coils of wire—with no external source of power, sitting on a table. According to European engineers who have examined it, a small feedback circuit provides energy to continuously motorize the discs, thus keeping them in motion. In 1989, an American new-energy researcher reported there were four of the energy converters in the commune, running continuously while generating ten to twelve kilowatts, in addition to two wind turbines powering special batteries, for powering houses, shops, and greenhouses.

The film goes on to state that nature is the greatest source of power and knowledge, "and it [nature] still conceals many secrets which are only revealed to those who approach [nature] . . . with the highest respect and responsibility." How do people understand these secrets? By experiencing silence and solitude, the commune says, stating that this is how Paul Baumann came upon the knowledge of how to build a new-energy machine.

This search for solitude is why Methernitha has been concerned with buying land—valleys, forests, mountains, or lakeshores—"where one could study nature, one's own being and the creator of all this universe in silence and concentration and without being disturbed." The film says that the public never understood this quest. Instead, onlookers interpreted Methernitha's seclusion as evidence that the group had something unsavory to hide. The truth, according to the film, is that the community has to go to great efforts to remain undisturbed in order to accomplish its goals.

The community refuses to reveal the secrets of the Thesta-Distatika, and stopped letting outsiders see it in the early 1990s. While the commune members go ahead with their lives, energy researchers elsewhere try to replicate the energy converter, looking for hints from European engineers who have actually visited the Swiss commune. Don Kelly of Florida, who cofounded the Space Energy Association USA in 1990, is a key networker in the information-sharing that goes on, in which tips on how to build the Thesta-Distatika are quickly transmitted by fax. Kelly and others are determined to figure out the puzzle for themselves. These researchers believe that the planet needs a clean, cheap, and abundant source of energy to remedy environmental and economic problems. The commune agrees with that goal, but does not want to be responsible for the possibility that some may turn space-energy technology into a weapon of war.

Too Dangerous to Disclose?

In 1986, Danish engineer Albert Hauser and two companions braved the February cold to visit Methernitha. As in earlier visits by outsiders, commune members were more willing to answer questions about their lifestyle than about the technical details of their machine. They described themselves as living like early Christians, in a community with its own school, machine factory, and market garden, and even a film studio. "They didn't want to publish the developed technic [technologies], which have taken approximately twenty-five years to develop," Hauser wrote, "Especially because they were afraid of the possibility of misuse from the weapon-industry." This statement is confirmed by Methernitha's film, in which the commune says that the world is not spiritually ready to learn about the Thesta-Distatika, and that the knowledge would be used for destruction. The film does not elaborate on their fears.

A representative of Methernitha wrote to Don Kelly in 1988, saying in effect that the group refuses to give away the secret of their invention because it might then be responsible for the consequences. The commune said that to surprise today's population with such a device would be like pouring oil on a fire: "What mankind needs is peace—peace of mind to start with—to have a chance to find nature and God and not further technological support in his [striving] for pleasure which would rather drown him in an ocean of noise, over-action, and pollution." The Thesta-Distatika may have also drawn the interest of NASA. One scientist wrote in a private letter to one of the Thesta-Distatika networkers that NASA officials had offered Methernitha a considerable sum of money for the device, but that the community declined the offer.

JAMES GRIGGS'S HYDROSONIC PUMP

Like Methernitha, James Griggs of Georgia is already experiencing the energy revolution. But he is not at all reclusive. He has invented a heating unit that, to his initial surprise, puts out much more energy than it takes in by creating shock waves in water. New-energy theorists want to solve the mystery of how the process works, but Griggs's customers just want to solve their down-to-earth problems. Since he and his partner together have put more than a million dollars into the project, Griggs naturally focuses on getting his product to his customers.

Putting Shock Waves to Practical Use

The incident that started Griggs, an electrical engineer with fifteen years of experience as as energy-efficiency consultant, on the road to invention occurred during a routine job in 1987. He was checking energy consumption in a commercial building when he noticed that water pipes on the way to a boiler were unusually warm. The company's engineer told Griggs that the heat was caused by water hammer, and was nothing of interest.

Water hammer, also called cavitation, is a condition that causes pipes to knock noisily. Sometimes, when a fluid is moving quickly through a pipe, pressure drops in part of the pipe, and bubbles form. These bubbles collapse when they are carried to areas of higher pressure, creating shock waves that collide with the inside of the pipe. It is considered a problem because the impact of the shock waves can pit the metal, damaging the pipe.

Griggs wondered if the problem of cavitation could be turned into a benefit—the production of heat—without the metal-pitting obstacle. He asked himself, "What if we used shock-wave technology to heat water?"

He worked on the idea at home, in his spare time, and eventually came up with a workable design. His pump has a cylindrical rotor that fits quite closely within a steel case. As the rotor spins, ordinary water is forced through the shallow space between the rotor and the case. The rotor is designed to create turbulence in the narrow gap, which heats the water and thus creates steam.

The surprise came in 1988, when a testing expert found that the heat energy put out by Griggs's Hydrosonic Pump was more than the electric energy input by a wide margin—the pump put out 10 to 30 percent more energy than was needed to turn the rotor.

The Hydrosonic Pump Hits the Market

As an energy consultant, Griggs found opportunities to test his experimental system in factories where a water-heating process was needed. Then, satisfied that his idea was sound, he took a chance. In 1990, he closed down his energy-consulting business and started up Hydrodynamics Incorporated. He funded the company out of his own pocket until 1993, when he took on a partner.

At first, sales were slow, although Griggs continued to experiment, building more than 700 different rotor designs. In 1992, the company sold a pump to a fire station in Albany, Georgia—a pump that is still producing excess energy. And hope came to the struggling company the next year, when *Popular Science* ran a cover story on cold fusion (see Chapter 8). Griggs read about these excess-heat experiments and felt that perhaps science could explain the results he was getting.

Through contacts with a network of cold-fusion researchers, and at new-energy conferences in the United States and Russia, Griggs began to learn about an effect that seemed to be related to his device. This effect, called sonoluminescence, occurs when ultrasound hits liquid molecules, causing them to give off light. In many tests, the Hydrosonic Pump had been seen to give off blue-colored steam. Griggs used his newfound knowledge to improve the pump's efficiency.

High efficiency is one of the pump's main strong points, along with low maintenance. All electric boilers start out at 100-percent efficiency, but their efficiency declines once minerals in the water start to clog the mechanism. This means that standard boilers need to be cleaned. The Hydrosonic Pump is different. "Nothing builds up," Griggs says. "It's self-cleaning."

From time to time, Griggs had seen another unusual effect within his pump—a barely discernable melting that occurred on the outside of the rotor on several occasions. This would have required temperatures of about 1,200 degrees Fahrenheit, much greater than any temperature that steam could produce without being under extremely high pressure. Even more amazing was that the microscopic bits of melted material rewelded themselves to the rotor. Such welding would require even higher temperatures—about 4,000 degrees Fahrenheit. It is obvious that no mere water hammer is at work within the Hydrosonic Pump, and its mysteries intrigue new-energy researchers.

The new-energy scientists might have been fascinated with Griggs's device, but the orthodox science community either remained skeptical or scoffed. More than 100 engineers showed up to do tests, and none of them denied that the pump produced steam and heat. But they all said, "You are making an error somewhere in your calculations," even as they were doing the tests.

But there is new interest in the Hydrosonic Pump from the scientific community. The civil engineering department at Georgia Institute of Technology is studying the pump in an effort to find out where the excess energy is coming from. And the local utility company, Georgia Power, has shown interest in displaying the pump in its new-technology development center in Atlanta.

Hydrodynamics is moving in new directions. The company has teamed up with a company in Florida to adapt the pump so that it heats synthetic oil instead of water. Such a unit would eliminate the fire hazard that is present when oil is used in a conventional gas-fired or electric boiler, since the Hydrosonic Pump doesn't use combustion. Griggs says that the pump can also be adapted to other uses, from milk pasteurization to pollution control, and may be useful in the space program—"if you put something out in space, you want [it] to be combustionless, if possible."

In the next part, we will look at not only how and why the energy innovators have been harassed, but at what the coming energy revolution will mean for all of us.

Part IV
The Energy Revolution—
Potential Amid
the Problems

So far, we've met a number of scientists who are working toward nonpolluting ways to generate abundant power—nearly free energy. We will now consider how these inventions could affect our lives. In order to picture what an energy revolution might mean to society, we can start by recalling our reactions to major changes on the personal level—divorce, losing a job, winning a lottery—events that shake up our lives. The emotional reactions of individuals facing such events can only hint at the mix of confusion, fear, creativity, and rejoicing that could accompany a global switchover to a world economy based on clean energy. The world has never experienced a reshuffling on such a scale, and the resistance to this change will be in proportion to the size of the entity that must learn to deal with such change. In this case, the entity is colossal—a world economy based on fossil fuels. It is a formidable entity, with an outlook weighted toward military-industrial alliances.

The pace of change will not please everyone. Defenders of the status quo are known for delaying or even blocking change, while impatient environmentalists may want to dance in the streets to celebrate the sudden death of King Oil. These opposites have clashed throughout the twentieth century, usually in incidents unseen by television news cameras. We will first peek into the laboratories where some of these clashes have taken place. We will then look at the impact new-energy technologies could have on society, and ask, "Where do we go from here?"

13
Harassing the Energy Innovators

The alphabet soup of [government] agencies . . . born of World War II and superpower confrontations, spends $35 billion or more each year of the taxpayer's money to sustain an unaccountable suppressive force that is now abusing its power.

—Brian O'Leary,
Physicist

It mystifies me why free energy is suppressed.

—Paul LaViolette,
Systems Engineer

H arassment of some sort or another has been a stumbling block encountered by most of the inventors whose stories are told in this book. Some of it has been as subtle as a turned back, while some has been as overt as gunfire. Most of it has fallen somewhere in between. In this chapter, we will take a closer look at what kinds of harassment these new-energy innovators have had to endure, and why.

THE FORMS OF HARASSMENT

The inventors we have met so far have run into different types of harassment: break-ins and destroyed equipment, business troubles, funding cutoffs, government pressure, hostility from the science establishment, lack of interest, violence and threats of violence.

It must be said that, in some cases, inventors have created their own problems. Nikola Tesla and John Keely made some questionable business decisions, and Lester Hendershot and Floyd Sweet deliberately misled others in the belief that their ideas might be stolen. But I believe that the human frailties shown by some of the new-energy innovators are vastly outweighed by the deliberate attempts made by others to impede research and development in this area. And the inventors we have met so far are only a few of those who have reported various forms of harassment.

THREATS AND INTIMIDATION: OTHER EXAMPLES OF INVENTOR HARASSMENT

In this section, we will meet some other inventors who have been harassed. Some of this harassment has been heavy-handed, such as jailings and break-ins. But some of it has been much more subtle.

When the Police Come Calling

In 1992, the fiery Austrian-Bulgarian physicist Stefan Marinov was outraged. His business associate, Jurgen Sievers, had been jailed for four months in Cologne, Germany, without a trial on a charge of investment fraud. Marinov went to the top. He wrote to the president of the German Federal Republic, Dr. Richard Von Weizäcker, demanding Sievers's release. A resounding silence greeted Marinov's six-page letter, so he sent the letter to friends in the United States. Here is the story he tells.

Six armed policemen burst into the home of Jurgen and Gerda Sievers at 8 A.M. on May 19, 1992. Sievers headed a company named Becocraft, for which Marinov acted as a science consultant. "After a many-hour investigation, at which even the beds were ransacked, all papers of the company were confiscated. A month later, Mr. Sievers was arrested on the street when, as in a stupid detective story, a police car blocked his way."

Cologne police then informed Sievers about the fraud charge. However, none of Becocraft's investors made such a charge, although invited to do so by the prosecuting attorney. Instead, the accuser—and the only plaintiff—was the utility company of Cologne. The utility company's stand was that Becocraft collected money to research and build machines that are free-energy devices, which the utility says are impossible to create. Therefore, Becocraft was involved in fraud.

Professor Marinov told it as he saw it: "Information has come to the ears of the gentlemen from the Utility Company Cologne that if such machines appear on the market, people would cut their wires leading to the electric stations and their cars would no more stop at the petrol stations." The physicist imagined the reaction: "'Any research and development of such machines must be nipped in the bud.'"

Marinov fought back. Half of his letter to the German president consisted of technical information intended to show that standard physics is based on incomplete or wrong teachings. Becocraft's free-energy plans are not fraudulent, Marinov said. Rather, current physics is incorrect in its denial of the possibility of devices that produce more energy than they consume.

"With this fraudulent science, the companies, research institutes, and universities steal millions of marks from the pockets of the German taxpayers . . . as with their energy sources they ruin the whole world. And these defrauders have thrown Mr. Sievers in the prison, because, with the money which some people voluntarily invested, he intended to save our world."

Europe has a closetful of stories about vested interests that have suppressed discoveries. Josef Hasslberger, a researcher in Rome, writes, "Tomorrow's technology has been invented, hundreds . . . of times, but each time it has been locked away in the vaults of the conventional energy cartels. If the inventor could not be bought, he was inactivated in other ways." He adds, "The amount of capital and thereby power involved is obviously too great to allow our fossil energy to be put aside too quickly."

Breaking and Entering

Break-ins are another form of overt harassment. T. Townsend Brown is a good example. Brown was an inventor working on new propulsion concepts that linked electricity and gravity; his antigravity devices literally flew against the known physics of his time. His first break-in experience occurred in 1945, when he was a retired United States Navy officer and consultant working at the Pearl Harbor, Hawaii, naval shipyard. He tried to interest his superiors in his work, and demonstrated his strange flying discs to a top military officer.

At the demonstration, Brown's colleagues treated his unusual discovery lightly. But someone must have taken it seriously, because when he returned to his room, it had been broken into and his notebooks taken. A day later, military spokesmen came to

Brown to say they had his work and would return it. After two more days passed, they returned his notebooks and said that the military was not interested.

Brown was disgruntled, but would not give up. Back on the mainland, in Cleveland, Ohio, he built a demonstration project called Project Winterhaven in 1952. The next year, he flew his discs for Air Force officials and aerospace industry representatives. The saucers whizzed around the course so quickly that the test results were stamped "classified" by the United States government. But again, it seemed that no one was interested.

After time spent in France, where yet another funding source had fizzled out, Brown returned to the United States, where he again ran into closed doors at the Pentagon. Even an old classmate from officers' candidate school, an admiral, tried to discourage him. "Don't take this work any further; drop it," was the essence of the admiral's advice.

Brown did not heed the warning. He moved to California, where he practically went door-to-door in Los Angeles to try to rouse some interest in his work within the aerospace industry. One day he returned to his laboratory to find it had been broken into and trashed, and that many of his belongings were missing. According to one of his associates, he also became the subject of nasty, character-discrediting rumors.

Brown went into semiseclusion in the 1960s, and died, as someone who knew him put it, "a deeply disappointed man." It wasn't until after his death that new-energy researchers uncovered evidence that the military may have been working on a project based on similar principles.

Astrophysicist Adam Trombly of Colorado was another break-in victim. Trombly, best known for an environmental public-information effort called Project Earth, worked openly in the space-energy field until the late 1980s. As a worldly public speaker and scientific networker, he had known what went on at government laboratories and other high-security places. But even Trombly has had the shock of seeing his laboratory violated—by professionals. He told a new-energy symposium in 1983 that he had bought the best alarm system available:

> If anyone breaks in, people are supposed to listen in and call the police. Well, we've been broken into three times and their computers seem to fail every time; I have the printouts of their

computer being down for forty-five minutes one time, twenty-five minutes the next time, fifteen minutes the last time. Whoever it was, was good, because they reconstructed the lock every time so you could not lock the door.

Tuning Out and Telling Tales

Although opposition to energy breakthroughs can take rude forms, other incidents are more subtle. For an example of a low-key but effective form of opposition—just ignoring a breakthrough—consider an experience described by the electronics editor at *Machine Design* magazine.

John Gyorki, in a 1989 editorial, addressed middle-aged car enthusiasts who remembered stories from the 1960s of fabled 100-mile-per-gallon carburetors. Popular folklore said that the devices were suppressed by oil monopolies, who always bought the patents to gas-saving devices so nobody would use the devices.

Like most well-educated technical people, Gyorki believed that such stories were nonsense and that automobile companies were doing their best for the consumer. "At least, that's what I thought. Then I got involved in a project that gave me second thoughts about my faith in corporate America."

As an engineering manager for an automotive supply company, Gyorki was asked to test a device designed to increase mileage and reduce pollution by producing a fine mist of gasoline within the engine, leading to more complete combustion. It was developed by an independent inventor who had turned it over to the testing and marketing experts at Gyorki's company. The inventor called the simple gadget a Petroleum Economy Device (PED).

"I approached the test with some skepticism," said Gyorki. However, the PED improved fuel economy by 25 percent and cut exhaust emissions by 85 percent.

The company rushed to Detroit, expecting a warm welcome. Midlevel engineers at the big automobile companies shared their enthusiasm, but the top executives turned their backs. They did not want to hear about a revolutionary improvement in mileage and emissions reduction.

A disillusioned Gyorki wrote that outdated technology "continued to rule for the next decade," ending only with the advent of fuel injection in the 1980s.

More subtle than ignoring a radical new technology, and much less honest, is the technique of disinformation—planting false information in order to hide facts from the public. One scientist with contacts in international circles says:

> They may have spent more money on disinformation than on developing technologies. It works; they [he declined to spell out who "they" are] learned that if you give the public disinformation, people become confused and passive. . . . It becomes a kind of cognitive dissonance; people tune out.
>
> Obviously, when you're dealing with a sensitive area, the government will protect its interests. We would be naive [to think there are not agencies doing this]. The only way to make legitimate information *inert*, less viable, is to embellish it . . . color it with all kinds of Looney Tunes. . . . Give contradictory information. So no matter how much the public might be interested in the topic, it creates enough confusion that people tend to get passive.

Checking Out the Competition

Would-be spies—from either private corporations or governmental agencies—either irritate or amuse different new-energy researchers, depending on the level of annoyance. Inventor George Wiseman, whom we met in Chapter 10, chuckles when he hears a click on his telephone line: "Any spooks out there listening to this tape you're making—have fun." Wiseman says that he has visited his telephone company to report the clicks and has been told by a company representative that someone was tapping into his phone line.

Wiseman and his friends also laugh about an incident in the early 1990s, when several men drove to where Wiseman lives in the forested mountains of southern British Columbia, parked near Wiseman's "Eagle Research" sign by the highway, and watched the property through binoculars for hours. The only calling card they left was a trail of cigarette butts.

Speakers at new-energy conferences, such as veteran inventor and publisher Ken MacNeill of Georgia, occasionally acknowledge the presence of incognito government and corporate representatives. "Some of the oil companies have a squad of people who go around and check out all these devices. We probably have somebody in the audience right now," MacNeill said at a meeting in 1994.

Paul LaViolette, Ph.D., of Vermont drew nervous laughter from the audience at a 1991 conference in Boston when he referred to "what I call the `free-energy police,' who are in attendance here." He added, "There are a number of people with the idea that there's some sort of unwritten law that if such a [free-energy] device is demonstrated, then it is impounded. . . . I understand there is a whole warehouse full of free-energy devices."

Another panelist released the tension in the room somewhat by joking, "Any Free Energy Police in the audience, please stand up." LaViolette, however, wanted free discussion of a troubling issue, and he addressed those nameless men whom he believes are just doing their jobs: "We're not blaming you. But I think this is important. If there is a law against [free-energy technology], can we be informed about it?" There was no reply.

THE BIGGEST STUMBLING BLOCK: THE PATENT OFFICE

In the United States during the last four decades, thousands of inventors have had the rude shock of sending in a patent application, only to see their work snatched out of their control when a government agent decreed that the work was classified. (See "Keeping Inventors Quiet" on page 163.)

An Inventor is Silenced

Adam Trombly knows about the Secrecy Act. In the early 1980s, Trombly and another young scientist, Joseph Kahn, Ph.D., naively believed that the "experts" would welcome their space-energy invention. (See Part II for more information on space energy.) However, when Trombly and Kahn applied for a patent, the United States Patent Office notified the Department of Defense. Instead of congratulations, Trombly and Kahn received a secrecy order. They were ordered not to talk about their invention to anybody, not to write about it, and even to stop working on it. They certainly couldn't tell the media.

In 1990, I asked a United Nations official about Trombly. His reply was, "Yes, I know about Adam Trombly's work. But there are powers who don't want his technology to come out."

Trombly echoes MacNeill by telling new-energy researchers to bring their technologies into the free enterprise system skillfully, without necessarily applying for patents:

I've become really tired of not sharing information. I'm under contract not to say this and not to say that, and the government sends you a letter saying you'd better not talk about this. It's hard to make progress that way. Let's form our own communication network, completely independent of any great corporations.

This plan, voiced in 1983, did evolve as use of computer networks spread around the world. Many new-energy researchers have decided against patenting their inventions, vowing instead to tell the world about their work through the Internet when they believe the time is right.

Lobbies, Legislation, and the Patent Office

In American mythology, "Yankee ingenuity" is rewarded. A clever person invents a better mousetrap and the world beats a path to his or her door. But in this century, that myth has changed. In the new mythology, a clever young person goes to work for a corporation, improves on its products, and is listed on a patent that is assigned to the corporation. The employee may be rewarded with a bigger paycheck—but in nothing near the amount of profit that the company sees from the improved product.

The lone inventor who tries to buck this system runs into a variety of problems. Corporate lobbies in Washington push for changes in the patent process—legislated changes that shut out independent inventors because it becomes expensive to get and keep a patent. This push doesn't just come from American corporations. Don Costar, founder of the Nevada Inventors Association, says that "[American] big business and foreign interests lobbies are combined. What they lobby for are patent law changes for global marketing benefits."

For example, under patent rules in the United States for the past two centuries, a patent is given to the first person to invent an item. In contrast, foreign patents are on a first-to-file basis. As a result, according to Costar, "The little company gets pushed out and stepped on by big companies who have a staff of lawyers who can draft a patent application overnight."

Inventors' advocates say that the United States patent system is being dismantled in other ways, such as through attempts to have patent information published just eighteen months after the patent is

Keeping Inventors Quiet

If you were an inventor trying to patent an important new-energy discovery, you might receive a secrecy order along the lines of the one reproduced here. According to information obtained under the Freedom of Information Act by the Federation of American Scientists, the Pentagon placed 774 patent applications under secrecy orders in 1991—up from 290 in 1979—and 506 of these orders were imposed on inventions by private companies. The government has standing gag orders on several thousand inventions. The following order, issued in the 1980s, was obtained by inventor Ken MacNeill of Georgia and revealed in 1983.

SECRECY ORDER
(Title 35, United States Code [1952], sections 181–188)

NOTICE: *To the applicant above named, his heirs, and any and all his assignees, attorneys and agents, hereinafter designated principals.*

You are hereby notified that your application as above identified has been found to contain subject matter, the unauthorized disclosure of which might be detrimental to the national security, and you are ordered in nowise to publish or disclose the invention or any material information with respect thereto, including hitherto unpublished details of the subject matter of said application, in any way to any person not cognizant of the invention prior to the date of the order, including any employee of the principals, but to keep the same secret except by written consent first obtained of the Commissioner of Patents, under the penalties of 35 U.S.C. [1952] 182, 186.

Any other application already filed or hereafter filed which contains any significant part of the subject matter of the above identified application falls within the scope of this order. If such other application does not stand under a secrecy order, it and the common subject matter should be brought to the attention of the Security Group, Licensing and Review, Patent Office.

If, prior to the issuance of the secrecy order, any significant part of the subject matter has been revealed to any person, the principals shall promptly inform such person of the secrecy order and the penalties for improper disclosure. However, if such part of the subject matter was disclosed to any person in a foreign country or foreign national in the U.S., the principals shall not inform such person of the secrecy order, but instead shall promptly furnish to the Commissioner of Patents the following information to the extent not already furnished: date of disclosure, name and address of the disclosee, identification of such part; and any authorization by a U.S. Government agency to export such part. If the subject matter is included in any foreign patent application or patent this should be identified. The principals shall comply with any related instructions of the Commissioner.

This order shall not be construed in any way to mean that the Government has adopted or contemplates adoption of the alleged invention disclosed in this application; nor is it any indication of the value of such invention.

At the conference where he revealed the secrecy order, MacNeill advised inventors of new-energy devices to go public: "Get the information or the device out there to enough people that they could not stop you."

filed. Such a system gives the edge to big companies with the resources to quickly turn a patent into a product. This isn't surprising, since patent-related legislation is "usually drafted by the lawyers of these big multinational companies," according to Costar.

The size of patent-filing fees is also a problem. Costar says that fees have jumped as a result of a Patent Office push to computerize its operations, and that inventors object to being saddled with these costs. It can cost an inventor several thousand dollars just to file for a patent, and additional money to maintain it as the years go by.

However, inventors now have a champion. A philanthropist named Steven Shore is funding a lobby to speak up for independent inventors, whom Costar estimates to number more than 4 million Americans. The Alliance for American Innovation is teaming up with a number of independent inventors' groups.

PARANOIA OR GENUINE FEAR?

New-energy researchers who have quietly worked for decades on their inventions without being threatened or suppressed in any way—other than being unfunded—find it hard to believe the ubiquitous scare stories of sinister "men in black suits" who purport to be from a high-level political or economic power. One story I was told involved a man who was intimidated just for helping others involved in new-energy research. Here, without the names and places, is part of a 1993 telephone interview:

Inventor: I told you about the . . . person who we were sharing some of our work with. . . . He was a military officer who was picked up and detained without them even telling his superiors where he was. They held him for twenty-four hours, totally questioning him about who we are and what we're doing.

Manning: Who are "they"?

Inventor (after a stunned silence, apparently surprised at being asked): They didn't identify themselves. They were the "men in black," or your-tax-dollars-at-work, or whatever. We tried to find out why he wouldn't associate with us anymore, and he finally told us, and then he was transferred out of the area, but we stopped all of our work in the city and moved it to another state after that. There are wild things that can happen at any time. It can be totally beyond your control.

While that last statement may be an exaggeration, the fact remains that I have heard too many firsthand stories, including some that occurred very recently, to believe that they could all be fantasy. Inventors have reported laboratories or home workshops being broken into after the inventor received some attention for an energy-related discovery. The damage descriptions indicate the presence of an intruder or intruders who were not there to steal anything, but just to upset the inventor.

One man who wants to remain anonymous, the inventor of a small fuel cell that runs on seawater, told me that two men visited his laboratory. While one was distracting him in conversation in another room, the other slipped into the workshop and apparently sprayed something on the wall. Later, when the building was unoccupied, a fire raced up the wall and destroyed part of the inventor's equipment.

A busy professional and part-time free-energy researcher in Texas says that he has been followed by "those clowns" who never introduce themselves. Several times, his associate has returned home to be harshly confronted with ransacked living quarters.

As other inventors tell it, the visitors who have threatened and intimidated lone researchers into silence do not introduce themselves by their given names. Nor do they leave calling cards. It is almost impossible, therefore, to verify many of the stories about suppression.

A New-Energy Nightmare Story

What happened to nuclear physicist Paul Maurice Brown is an unfortunate example of what can happen to the independent researcher. Brown began researching magnetic energy devices in 1978, when he was a college student. Over the years, he heard what he describes as nightmare stories about inventors who made a breakthrough and then were persecuted, harassed, or even killed. He believed these stories to be a result of inventor's paranoia, a belief confirmed by encounters with several inventors with wild imaginations who seemed to be their own worst enemies.

Brown developed a novel method for converting natural radioactive decay directly into electricity in the form of a battery. In February 1987, the proud inventor and his associates at a private research company in Boise, Idaho, decided it was time to make a public announcement.

A series of traumatic events followed. The state departments of health and finance filed complaints against both the company and Brown. His license for handling radioactive materials was suspended. He began to receive anonymous threats, such as "We will bulldoze your home with your family in it."

Relocating the company to Portland, Oregon, did not stop the troubles. Despite the fact that a 1988 *Fortune* magazine article commented favorably on the nuclear-battery venture, securities fraud charges were filed against both Brown and his company. Oregon's state finance department investigated him, as did the Internal Revenue Service and the Securities and Exchange Commission.

After meeting each challenge, Brown redoubled his efforts to develop his technology. But events worsened. His young wife was assaulted. Even in their home they did not feel safe; it was robbed three times and vandalized on four other occasions. Brown was twice accused of drug manufacturing and, eventually, lost control of his company. The Browns also lost their home.

Finally, the pipe-bombing of his mother's car in the early 1990s drove Brown to become a recluse. "I understand now why inventors drop out of society," he said in a 1991 open letter to other new-energy researchers. His advice to them? "Keep a low profile until you have completed your endeavor, be selective in choosing your business partners, protect yourself and your family, and know that the nightmare stories are true."

THE REASONS FOR HARASSMENT

At a 1988 meeting of the International Tesla Society, Adam Trombly expressed the hope that the suppression stories he had heard would get out into the open: "We are hoping that more people will come out of the woodwork."

What can an inventor do? Trombly suggests two options. One would be for the inventor to remain completely anonymous—find an investor who could tuck the operation away in a remote place, and stay unknown. The other option is to "tell everybody what you're doing and then nobody will get at you because you aren't keeping any secrets. . . . Whatever you do, be careful. It's a real world out there."

Toby Grotz of Colorado is an engineer and organizer of new-energy organizations, including the International Tesla Society and the Institute for New Energy. He says that he does not believe in suppression "as an organized conspiracy. I think the suppression is occurring within individuals—our resistance to change." He adds:

> The collective consciousness hasn't decided that it's okay to make the quantum leap to free energy. . . . The collective consciousness did decide it was okay to control fire, the wheel, steam and gas engines, electric motors, nuclear power—all these little technological stairstep jumps through our evolution are a result of collective consciousness saying, "Okay, it's time to do this. Now we can advance."

British author John Davidson concurs: "Many of those finding themselves in a struggle against the `system' have been prey to paranoid feelings of victimization and suppression, when really there had been no plot to suppress their work, but only the subconscious inertia of `established opinion.'"

I also believe that there is no organized conspiracy. There is the factor of greed in human societies, but often suppression arises from the human fear of change, our fear of the unknown. But while not many know about the possibility of a new-energy world now, that could change very quickly. Moray King, the space-energy theorist we met in Chapter 4, says that as the worldview does shift towards the acceptance of new energy, "the special interests that went into suppressing the discovery will then produce enormous investment capital to develop it—if you can't lick them, join them." At this point, the space-energy industry will grow as fast as the computer industry, King says, and will offer many new opportunities.

No matter how much industries change, the independent inventor will always exist, because becoming an inventor is not a choice that is made with cold reasoning. It is more often a compulsion than a lucrative pastime. It is often a hobby that strains the patience of family members, empties the family savings account, and fills the house with tools, sawdust, reference books, and pieces of metal. If the laws of the land were not so blatantly in favor of the corporation, and harassment in all its forms would disappear, perhaps the beleaguered independent inventor would have a better chance of succeeding.

In the next chapter, we will look at the economic implications of free energy.

14
Society and a New-Energy Economy

The public is a lot more powerful than special interest groups. But the public is asleep.

—John O'Malley Bockris,
Physicist

Innovators are moving toward extremely low-cost clean energy technologies—much faster than political leaders in North America are preparing society for the resulting economic shock.

In *Road to 2012*, a report for the United States Coast Guard, futurist John L. Peterson warns about possible human suffering that could result from a changeover in world energy economics. Peterson says that our current system of fossil-fuel energy will be rendered obsolete, replaced by a system in which the new-energy sources examined in this book will become society's main power supply. He says:

> On the one hand, great hope would attend this new way of solving huge global problems. A new era would loom on the horizon. On the other hand, shifting to the new mode would not be easy for those who cannot change easily and quickly. This would produce great despair for many.

What might the future hold? In this chapter, we will first look at the pitfalls and possibilities of a changeover to a new-energy economy. We will then look at the forces of opposition, and how they may be overcome. We will see whether or not the changeover will occur, and what a transition period might look like. Finally, we'll see what will be required for us to move toward a life after oil.

NEW-ENERGY PITFALLS AND POSSIBILITIES

Peter Lindemann of New Mexico—author, inventor, and long-time researcher of energy alternatives—says that people must discuss how a new-energy revolution would affect society. He sees the technical know-how as being at a point where such a revolution could happen rapidly—probably within a decade. But, he says, "unless something really changes on the social or political or economic level, the technology is irrelevant; it will not be allowed to happen."

What are the roadblocks between our old-energy present and our new-energy future? Many of these roadblocks are built in the offices and boardrooms of big players in the current economy. People familiar with the ways of government say that a highly placed employee in a federal energy- or invention-related agency will receive an offer of a well-paid future job from someone in industry, based on the employee's performance in blocking developments that would mean less profits for the industry and in generally maintaining the status quo.

Another roadblock is the push for corporate profits. This gets in the way of decisions that would help bring new-energy technologies to the marketplace. For example, the automobile industry continues to fight laws in California that call for a certain percentage of new vehicles to produce zero emissions, since it would be difficult to recoup the companies' increased production costs on something like an electric vehicle because of the need to retool their assembly lines.

One of the most important roadblocks to progress is the lack of public awareness about possible new-energy sources. Unless the public knows about these sources, there will be no public pressure exerted on institutions both public and private to welcome them. Lack of knowledge about this subject is widespread; even government officials who shun corporate temptation are unaware of new-energy possibilities. And until now, peer pressure toward conformity among scientists and journalists has worked against education on this topic. Resistance to change is the underlying factor.

Other roadblocks are less visible. Wall Street provides the start-up money for oil and nuclear energy megaprojects, and depends on the continued flow of interest payments from such investments. The power of the financiers should not be underestimated. If they decided to stop loaning money for oil tankers, dams, and nuclear plants, such projects would not be built.

Another roadblock is the way the government itself is partially financed by energy. A portion of government revenue in the United States and in other countries comes from energy taxes. For example, in 1992, taxes on motor fuel brought in $22.25 billion to state governments. If the public suddenly used 20 percent less fuel, there would be less money flowing into government coffers.

A rapid changeover to a new-energy economy could also derail large numbers of jobs. For example, a large utility has an immense amount of money invested in capital equipment—hardware that is used to either make other goods or otherwise bring in income— and in bonds to cover debts, such as those incurred in building a nuclear power plant. If an invention suddenly made such a plant obsolete, the company could not decommission the plant and write off, or eliminate from their bookkeeping accounts, the dollars that have been committed to the nuclear path. This would bankrupt the whole utility, putting many people out of work.

There is a lot of money—and a lot of jobs—tied into the fossil-fuel economy. In 1991, the cost of energy in the United States was $891.1 billion, or 15.6 percent of the gross national product. When you add to this the number of jobs related to fossil fuel in the rest of the world, it becomes obvious why a fast changeover to new sources of energy could wreak havoc in the job market.

At the same time, the dark cloud of job loss has a silver lining of job creation. Lindemann says, "We have to take everything we have now, dismantle it, and replace it with something that's going to work for a future that's sustainable and that won't poison us. There's tons of work to be done; the idea that it's going to put everyone out of a job [in the long run] is totally ridiculous."

Humanity faces a test when new-energy technologies are accepted as a reality. The test will be in using them to enhance the quality of life and to clean up the earth, instead of using them to create weapons of destruction or more landfill mountains of consumer junk.

Some people believe that such jobs could give people more than just a means of support. H. D. Froning, Jr., of McDonnell Douglas Space Systems in California, is working on ways to use new-energy technologies in space exploration, believing such work will require the same sorts of technical breakthroughs that will be needed in energy production. He speculates that such technological advances would go far beyond meeting the bare essentials of life. People also need a sense of worth, and he envisions the creation of new occupations that would give useful work to expanding populations.

FORCES OF OPPOSITION

Opposition to new energy has come from several sources, as we've seen throughout this book. They include the oil companies and other large businesses, and surprisingly, a part of the environmental movement.

But what about the oil company executives? Lindemann notes that they realize their product supply is limited, and as a result, their companies have diversified. Today, they own coal and uranium mines, and produce plastics, fertilizers, and chemicals. The gigantic companies want to be in business, he says. They do not really care what they produce. This echoes the findings of distinguished physicist and energy researcher Harold Puthoff, who says he was told by oil executives that they would welcome a new source of energy, because they would make more profits by turning oil into plastics and pills than by selling oil as a fuel.

"I don't think the oil companies or the energy distribution companies are the problem," Lindemann says. "They just don't want to see things happen rapidly so that suddenly everybody is running on a heat pump in their backyard; that would only cause world depression."

In that case, what sector does he see as a major problem? In Lindemann's view, it is the media and a power elite that owns large magazines, newspapers, radio stations, and television channels, and has been known to manipulate public opinion on a new-energy discovery in order to maintain social stability. I believe that this elite thinks the populace might become angry and rebellious if average people realized that the problems of the fossil-fuel economy—high heating, electricity, and transportation costs, as well as pollution—are unnecessary. Lindemann says:

> An inventor gets suppressed. Sacrifices had to be made for the preservation of order. . . . I don't agree with their means of bringing this about, but I do agree with the end result [of social stability]. I don't know if they can accomplish it with

their low level of integrity, though. I think that if the entire project of social order was put out to the public, you'd get a lot more cooperation.

Who are "they"? Lindemann says that gigantic financial forces—the worldwide banking system and financial markets—work behind the scenes to mold the economy. "That's where change has to come from. In spite of the fact that I don't like they way they exploited us and don't like the fact that they made sure that power continued to concentrate in their hands." However, Lindemann's views differ from those of other new-energy advocates, who are impatient to get energy out of the hands of the big companies and into the hands of the people.

One trend that could affect the future of energy-technology changes is widespread disillusionment with corporate greed, greed that has resulted in destroyed ecosystems as well as corrupt political systems. "If the idealists who want a healthier world all pull out of the whole trend of where the power is moving us," Lindemann warns, "we're going to get the worst possible future." These people, many of whom look with fear and distrust at the economic trend away from national sovereignty and toward multinationalism, must stand up against the abuse of power and take part in shaping our future.

Where does the environmental movement fit into the new-energy scenario? One author, P. J. O'Rourke, ridicules environmentalists, making the case that some people actually want to live in apocalyptic times and would be uncomfortable with abundance. O'Rourke may not realize it, but he helps us understand why the new-energy movement gets little help from the environmental movement. He quotes long-time environmentalists Jeremy Rifkin, Amory Lovins, and Paul Ehrlich as stating that giving society abundant, bargain-priced, harmless energy would be the worst thing that could happen to the planet because of what people might do with it. They apparently would rather try to get a significant number of people—especially energy users in developed countries—to be energy-efficient, to tighten their belts, and to live with less energy for electricity and transportation. Most of the environmentalists whom I've worked with feel this way.

To be realistic, however, society will probably be getting low-cost, abundant energy whether the belt-tighteners like it or not. If Western environmental groups do not welcome new-energy technologies, the devices will be imported from the East (see Chapters 7 and 8 for examples of how new-energy ideas are welcomed in Asia). Would it not be wiser for environmentalists to relearn flexibility of thinking and help shape the direction of the coming energy revolution?

WILL WE OR WON'T WE?

To answer the question of whether or not a new-energy changeover will come, we should look at two views of change, and at what the transition period to a new-energy economy will involve.

Two Views of Change

I believe that a changeover to a new-energy economy is inevitable, but some people are not so sure. Let's look at both sides of the argument.

On the skeptical side, a retired magnetics-research technician in Cincinnati, Erwin Krieger, doubts that a "free energy" device would enter the consumer market in North America in the near future or "as far down the road as you care to look. It is quite probable that the military would snaffle it up first. And . . . probably with prohibitions against `unauthorized construction and use.'"

He adds:

> Then there's the economic impact. It's all very well blithely to prattle about buggy-whip makers trading their craft in for automobile accessories production; that parallel is far from realistic. What of workers at shipyards that make the huge supertankers that bring millions of tons of oil from here to there? What of the workers, technology and research world-wide, in the oil-production industry? What happens to the scientists and their research on solar energy, atomic energy? With the mining and vast infrastructure of coal production? Or oil, gas, and coal power[ed] plants?
>
> Although the numerous aspects of the energy business look disparate, they are, in fact the interconnected building blocks of one monolithic structure in which a crack anywhere would eventually collapse the whole.
>
> Had I a free-energy machine in my closet laboratory I would pack it up and hie me to some power-poor Andean or African country. . . . Of course, greed and politics being what

they are, sooner or later the country in question would consider exporting cheap power to neighbors and . . . need I continue? The introduction of a free-energy device—*that* would be an energy crisis!

On the side of the optimists is Bill Lawry of California, a successful entrepreneur who has helped to fund new-energy experiments. He says that if one of the inventors were to develop a reliable device, "it would be the most revolutionary event—magnificent and catastrophic all at once. In the long run the change [to new-energy technology] would be for the betterment, but in between there would be dislocation of a magnitude the world hasn't seen."

Lawry can see why someone would want to suppress the development of new-energy hardware, but says it would be an impossible job because there are too many gifted people who are determined to make it happen. He has wrestled with the question of what should be done if an energy device was perfected to the point where it was ready for mass manufacturing. Saying to the world, "Here it is!" at a big press conference is not his choice:

> I'm an entrepreneur, so my approach would be to go to four or five of the major companies—let them compete against each other—and say, "This is what we can do with this invention, and this is what it is going to do to your company. You've already got a large staff of engineers who can design the products and you have the facilities for manufacturing. Go to it."

Paying for a New-Energy Changeover

Of course, new-energy technology will not be free of cost. Physicist Hal Fox points out that engineering, materials, and other expenses have to be figured into projections for possible mass-produced new-energy technologies. His most optimistic prediction is that consumers could have clean power at costs at between one-tenth and one-third of current costs—about $1,200 a year for the average household in the United States.

How will the transition period be paid for, without causing large-scale hardship? I suggest to Peter Lindemann that the transition time could be compared to starting a new business, with everyone having to make some sacrifices until profits begin flowing.

"Your analogy is good," he says. "But it's easy to imagine what it looks like at the individual level. What's going to happen when an entire society has to do this, as well as a government whose entire tax structure is running on the energy used now? What laws have to be changed? As energy starts being used in different ways, how are the taxes shifted?"

What might a transition period look like? To kick-start it, taxes on old-energy technologies would have to be raised in order to fund research and development of new-energy technologies. Then, the fossil-fuel transportation, heating, and electrical generation hardware, as well as the nuclear power plants, could be phased out as new-energy hardware comes out of the factories.

Since this hardware would be fuel-less and durable, the old fuel-tax system would no longer work. But, rather than looking for another product or service that could be taxed, governments could make up the lost revenue by cutting back on spending on the defense-oriented parts of their economies, much of which has been rendered obsolete by the end of the Cold War. The United States Department of Defense alone spends billions of deficit-financed dollars annually. And since the United States government itself is a large consumer of energy, new-energy hardware could reduce its costs directly.

Private businesses can develop this new-energy hardware if the roadblocks we discussed earlier are removed. But they need the cooperation of federal and state governments, which must formulate a new energy policy that is wholeheartedly in support of the transition to clean, low-cost energy.

LIFE AFTER OIL:
MOVING TOWARD A NEW-ENERGY ECONOMY

Before we can consider a transition to a new world of energy, we must come to terms with our past behavior. Only then will we be able to see clearly into the future.

Acknowledging the Past

I believe that we cannot move forward in harmony without accepting responsibility for the past. These issues must be discussed, not to arouse negativity, but to instead hasten a collective acknowledgement of the problems we all face. This way, society can move on toward healing, in both attitudes and behavior.

A healing of attitudes may lead to a healing of planetary ecosystems. We must face the fact that we have all abused the earth. For example, my pickup truck burns gasoline. Therefore, it consumes oxygen and trails poisons out of its exhaust pipe. All the corporate feel-good advertisements for clean-running fuels that we read in the papers and see on television do not change the fact that internal-combustion engines abuse the planet.

But individual acknowledgement is not enough; there must be collective acknowledgement as well. John Hughes, M.D., a physician-psychologist and former political candidate for the British Columbia Green Party, says that most thinking people today are, at a subconscious level, grieving for their planet. He says that at a deep level, we know the deadly effects of excess radioactivity, of deforestation, and of chemical pollution on both the earth and its inhabitants. These unacknowledged feelings—about our collective actions and their results—sap our ability to act effectively. He suggests that we turn to each other for mutual support as we admit to our deep fears, and that one goal of the process is to be able to respond to the coming economic challenges effectively and with clear-thinking courage.

Facing the Future

Even though it will require an all-out effort by all of us, a changeover to a new-energy economy is desperately needed. Environmentalists thought that by now we would start turning off the energy industry's myriad spewing founts of pollution. Progress has been slow. One space-energy scientist voices his frustration: "I am becoming a revolutionary who feels angry about our shortsighted, suppressive, and ecologically destructive culture." Brian O'Leary, Ph.D., a co-founder of the International Association for New Science (IANS), adds, "I'm eager to help create those social structures that will facilitate a new worldview, one that will support a sustainable global future." For example, IANS has proposed the founding of an Academy for New Energy that would train scientists in new-energy theory and methods.

And what of the politics of energy, a politics that must change before the economics of energy can change? A number of writers in the energy field have addressed this subject.

Curtis Moore and Alan Miller, authors of *Green Gold*, make the point that the United States, with its creative edge and its resources, could win the energy race and get a large share of what O'Leary—

a former presidential advisor—estimates is the $2.1 trillion market for major energy technology coming globally over the next several decades. However, they say that an advantage of American society—its open political system—works to its disadvantage when oil and other industry lobbyists manipulate that system and use it to shut the doors on energy alternatives. Thus, the government reflects the needs of big business to a greater extent than those of other segments of society. Although Japan and Germany consider the needs of their industries, those governments "also maintain a clear vision of what their national interests require."

This is not a problem limited to the United States. Christopher Flavin and Nicholas Lenssen, authors of *Power Surge*, say that corporations and governments the world over "seem to be looking at the future through a rearview mirror." I think they're right.

But the switch to new energy will be like the switch from the horse to the car, or from the telegraph to the telephone, or from the radio to the television. It is unstoppable. Ultimately, the push of an ever-accumulating body of new-energy research will combine with the pull of an ever-increasing need for abundant, nonpolluting energy sources to create an irresistible demand for new-energy technology. Systems engineer Paul LaViolette gives voice to a widespread view when he says, "The whole [new energy] thing is growing so fast that suppression is not going to work, because it's going to break through anyway. Like any revolution, it can't be stopped."

In the next chapter, we look at how a new-energy society can emerge—and what you can do to help.

15

The Power Is
in Our Hands

Do you feel the crescendo? More and more inventors say they're nearly ready to go public.

—Gary Hawkins,
Inventor and entrepreneur

These are public issues that are of such importance, they need to be debated now—before control goes into the hands of the few for profit and power, rather than for the common good.

—Brian O'Leary,
Physicist

D o we want a new-energy future? I think if you ask most people, they would agree that we do. Are we willing to demand a new-energy future? That is the important question. If we do not demand a change for the better, then change—when it does eventually happen—may not be the change that we want. For example, even if new-energy inventions are developed in Japan or Korea for mass production (see Chapters 7 and 8), they may not be sold here if powerful economic interests remain opposed. As new-energy writer Michael Schuster says, "The desired end product is not necessarily a free gas pump in every household, but more a sense of empowerment."

One internationally connected American businessman speaks of a powerful factor in countering vested interests—the will of the people:

When the Soviet people got their hands on computers, faxes, and videos, information spread faster than the state could control [it]. The people demanded changes. That's what happened to the power of the Communist Party in the USSR. In the West, the same thing will happen to the energy cartels. Here, it comes from [the] Internet, you name it; the people are finding out about free energy and it's too late for the cartels to control it.

In this chapter, we will first look at what a new-energy world would look like. We will then see how far we've come down the road toward our future, and how far we have to go. Last, and most important, we will look at what each of us as individuals can do to help bring about the coming energy revolution.

THE IMPLICATIONS OF NEW ENERGY

What would a new-energy world look like? Think of the possibilities:

• Instead of fighting oil wars or financially supporting nuclear power plants, governments convert the plants to run on nonpolluting energy technologies and carry out large-scale cleanup projects. But most power is generated by privately owned devices, varying in size from a backyard generator to a plant big enough to light a city.

• Oceans, rivers, and forests are freed from the threat of further contamination by radioactive waste, oil spills, or acid rain.

• On highways and city streets, traffic hums along quietly without the roar of internal combustion engines. Even downtown, the breeze smells fresh and pure. Alongside the freeway, joggers can breathe in lungfuls of sparkling clean air.

• As fewer pollutants spew out of power-plant smokestacks, soils everywhere are cleansed and restored to health. This restoration is helped by the lack of heavy-metal fallout from gas and diesel engine exhaust. As a result, fruits, nuts, and vegetables grow anywhere, from greenbelts to inner-city backyards. Anyone with a rooftop can build a small greenhouse, heated in winter by fuel-less devices.

• Jet aircraft are converted to fly on water fuel with a technology that, at the same time, breaks up the existing chemical oxides that now contaminate the air. Thus, aircraft renew ozone in the upper atmosphere, instead of devouring it. This, in turn, reduces a host of problems, ranging from skin cancer among people to die-off among plants.

• A great number of constructive jobs emerge from a combination of abundant, low-cost energy and a gradual changeover to technologies that are in harmony with nature.

• The increased vitality of people who breathe oxygen-rich air, drink unpolluted water, and eat healthy food can result in an upward-spiralling surge of hope, creativity, and determination to solve humanity's problems.

What sort of devices could end up on the market? In addition to the inventions discussed in Parts II and III, there are other devices that are being developed:

• The space-energy silicon chip is a possibility. Adolf Zielinski of Wilmington, Delaware, a researcher with a background in high-tech business ventures, is working on a way to put space-energy technology (see Chapter 4) on a silicon chip. Such devices could run everything from cars to electric power plants to computers.

• Yasunori Takahashi of Japan has developed a space-energy motor that allowed a scooter to zip along at seventy miles an hour on a freeway. His Self Generating Motor uses the most powerful magnets known in the new-energy world.

• Magnets are used in another device that could power a home or run a car. Norm Wootan and Joel McClain of Texas use magnets and crystals in a device that puts out more power than it takes in. One new-energy writer says that the Magnetic Resonance Amplifier could allow the development of an electric car that generates power as it is driven.

• Solar energy may get a whole new look. Alvin Marks of Massachusetts is getting help from government researchers in developing a unique type of flexible solar-cell film. It would work like photosynthesis works in plants; sunlight would strike light-activated molecules in a conductive film, which would cause positive and negative electric charges to separate and flow in opposite directions. Such a device could be cranked down like a window awning in sunny weather to generate electricity.

• An old-fashioned energy source—the windmill—is becoming a new-energy technology. Bill Muller of British Columbia has developed a generator, based on a system of magnets and electric coils, that would allow a windmill to put out dramatically more power.

• Water is the basis for a new kind of fuel. Yull Brown of Australia has developed a special technique for breaking down water into unusual forms of its component parts—hydrogen and oxygen. The resulting gas can be used to do specialized welding, its current primary purpose, but can be adapted to run an automobile engine. Brown's gas also seems to reduce the amount of radioactive decay from nuclear waste materials, which means that it might be a way to decontaminate waste sites.

• Water is also the basis for an electrochemical reactor developed by Randell L. Mills of Pennsylvania. This reactor is unlike the cold fusion technology discussed in Chapter 8 in that it uses ordinary water as a source of heat for generating electricity. Other cold fusion-type devices use a gas instead of a liquid to produce power.

• The Energy Trimmer is already on the market for industrial and institutional customers. Melvin Cobb of California has invented a device that does not generate electricity in itself, but instead balances out the electric fields in large buildings, increasing efficiency by about 25 percent. Southern California Edison has approved the Energy Trimmer for its energy rebate program.

THE ROAD TO FREE ENERGY

Like a traveller who has just started on a long journey, we have just started down the road from a system built on fossil fuels to one built on new-energy sources. Let's look at both the road behind us and the road before us.

The Steps We've Taken

For all the problems the new-energy movement has encountered, there has been some progress. We will look at how the world of science fiction has prepared us for a new-energy future, and how the Internet is helping to turn the future's promise into reality.

Fiction Foreshadowed the Future

Until new-energy conferences started to be held on a regular basis in the 1980s, the only place where new-energy concerns received a serious hearing was the world of science fiction. From such novels as 1981 underground classic *Ecotopia Emerging* to television shows such as *Star Trek*, science fiction writers treated space energy and other new-energy topics as realistic possibilities, instead of as wild fantasies.

Popular culture helps to prepare the public for a change in their lives—in this case, a new-energy future. Many people do not believe in the reality of something they have not seen on the evening news, or bought at a store, or otherwise brought into their familiar surroundings. Familiarity with new energy through popular culture may help foster a quicker acceptance of new energy as a believable, workable entity.

A New Factor: The Internet

Can today's clean-energy proponents succeed where yesterday's could not? Today's renegades do have an advantage—the global electronic brain known as the Internet, on which new-energy information is transmitted at an ever-accelerating rate. The information revolution is marching hand in hand with the new-energy revolution. Many inventors make statements such as, "If anything happens to me, everything I know will be uploaded onto every computer network. I've made that provision."

The Internet is a global network of telephone wires, fiber-optic cables, and satellites through which a computer user can instantly connect with another user anywhere in the world. It allows lone researchers in various countries to exchange experiment results, research ideas, and—perhaps most important—encouragement and support. British author and new-energy researcher John Davidson says:

> This is a wave that many of us are jointly riding and which is simultaneously breaking, apparently independently, in all parts of the world. . . . Through the networking efforts of many people . . . the work is being drawn together . . . with such visible evidence of its reality that it can never again be brushed under the carpet by prejudice and vested interests.

Mark Hendershot is a good example. Mark—son of an inventor we met in Chapter 3, Lester Hendershot—knows firsthand how secrecy and suppression can devastate an inventor's life. Mark has a family of his own, and giving them a peaceful and healthy existence is his highest priority. He doesn't finance work on his father's generator through investors, or by by selling shares in a company. Instead, he sells information packets about his father's invention.

Like other inventors, Hendershot has decentralized the control of his own technological secrets. A scattered group of associates will dump his information onto the Internet if the signal is given. "I don't expect to get rich," he says. "I just want to get the information out to the people, for the sake of our grandchildren."

The Steps We Must Take

Obviously, we still have a long way to go. The road before us will require that we learn to accept the idea of abundant energy, despite the fact that we have been conditioned to believe in scarcity. We will also look at how women can help change the current scientific worldview, a view that will have to include energy technology in a framework, that of the larger web of life.

Digesting the Concept of Abundance

As we have seen by now, we do not live in a world of scarce resources when it comes to the potential power available to us. In that respect, we exist in a sea of plenty and the *politics of scarcity are illusory*. That thought takes a while to digest. Although leaders of the New Age movement—the philosophy that says we each create our own reality through the way we think—preach "think abundance and prosperity," society is conditioned to a worldview of scarcity and struggle.

It is no illusion that humans have devastated natural systems on this planet and have caused some resources to become scarce, and that we are running out of room for our garbage. But the main reason we can't reverse this situation is the belief that we cannot regain control. Vested interests—large corporations and government bureaucracies—are formidable forces indeed, but not as strong as millions of people who believe in the concept of abundance.

Women and New Energy

A changeover time challenges conventional wisdom, and the implications of changing to cheap-and-abundant energy are immense.

How can such a world be sensibly run by the old monopoly-oriented rules? Some thinkers suggest that answers will evolve more readily if the base of decision-making is widened to include women.

Will parity between men and women make a difference? One researcher, a man, notes that "participation by women is a lack in this [emerging energy technologies] field. . . . This may partly account for the over-emphasis on power [produced] . . . instead of [on] what technology does to living organisms."

One scientist says that girls outnumber boys in environmental clubs in the schools. When he wonders why, his wife points out that girls learn early on that "if you make a mess, you have to clean it up." And another scientist who was contemplating a speaking tour to promote a book on the hydrogen economy (see Chapter 9) said he would target women's groups because he feels women are closer than men to their protective feelings toward future generations.

Do women really have a subtle difference in their perspective of life, a difference that is needed now? Beverly Rubik, Ph.D., describes her experience as a woman studying college science in the late 1960s:

> Even the biological models were mechanical and lifeless. Where was the Nature that I knew and loved—the gentleness, the delicate balance, the complex and subtle relationships, the diverse beauty? Gradually I came to realize that these were not an important part of the conventional scientific worldview.

Rubik points out that the language and methods used in science are often brutal. Smashing atoms and killing organisms are the accepted route to learning. High-tech products, from bombs to medicine, are products of a mechanistic science, a way of seeing nature as a passive mechanical object separate from the world of human beings. Some people believe that there must be more of a balance between masculine and feminine worldviews in international decision-making. Although no cure-all, this balance could help turn public policy toward more life-oriented energy policies.

Energy Technology and the Web of Life

Debate over how to balance energy technology and the need to give individuals more power over their lives—how much electricity costs, whether or not it comes from a nonpolluting source—may come at a time when many people are reexamining the materialistic basis of science as currently taught. One long-time observer of the new-energy scene says an evolution in technology is not as important as an accompanying evolution in understanding, an evolution that will open our eyes to another dimension of our living universe. Perhaps, the wisdom to use technology responsibly will increase when enough people begin to comprehend the way that all life is interconnected. While scientists involved in this field study the physics of space energy—the measurable world of atoms and forces—some of them go beyond the formulas and equations to express an awe at the beauty of what they are working with.

We have a lot to learn about the web of life as we jump into the new-energy era. A study of the different ways in which new-energy and old-energy technologies affect living creatures may be a place to start. Viktor Schauberger envisioned a "living technology," and today's inventors are working on it. When enough people agree to take responsibility for learning and applying new life-enhancing energy knowledge, deserts may turn green and fresh breezes may blow down city streets. Inventor Adam Trombly says, "Great technologies alone are not going to save this planet. Great humanity is."

Some are calling for biologists to help judge and make decisions on new-energy development. The science of biology is very relevant to the energy field because of the possibility that life forms respond to subtle energy fields produced by unconventional hardware. For example, a United States Navy project in northern Wisconsin involves a long antenna laid along the ground to communicate with submarines. This antenna may be sending out a strange form of electromagnetism, since trees in the area are growing abnormally fast. While some may see this as a benefit, others do not—one health researcher says, "I don't want my children to grow abnormally fast." These are the sort of effects that must be examined as we move into a new-energy age.

FREE THE POWER

There are indeed signs that change is coming. This book contains only a sampling of the individuals who say they are on the verge of being able to provide revolutionary energy technologies. There are also risk-takers in the business sector who are willing to finance the development of these technologies, and whom some expect will advance the next wave of the energy revolution.

But vigilance is needed even as energy science changes. Humanity has been lulled into blind faith in its scientists and engineers throughout the twentieth century, and the public is only now beginning to wake up and see what gigantic mistakes those experts have made in energy megaprojects and atomic experiments. Perhaps both women and men, and biologists as well as economists and engineers, will insist on having a voice in deciding humanity's new directions for the twenty-first century. Brian O'Leary offers a vision of how to achieve such teamwork, which would include "good government in concert with industry":

> The challenge is to find the scale of funding that would ensure the orderly and rapid development of the best technologies, rather than stumble into the grips of secrecy and the control by the few. I believe the challenge can be successfully met through the power of positive visioning and goal-setting.

But what about the average person? What about people like us, people who will be profoundly affected by these massive changes? There are signs of public interest in energy issues, such as the renewed interest in energy pioneer Nikola Tesla. And there are ways for you to get involved, both in helping to decide public policy and in making energy decisions for yourself:

• Use the Resource List in this book to learn more about new-energy possibilities. There are magazines and newsletters aimed at different levels of technical knowledge.

• Get involved politically. When a local, state, or federal election comes up, find out what the candidates know about new energy, and what their commitment to promoting it is. You will want to know if they will steer money away from fossil-fuel and nuclear developments and toward new-energy research.

• Don't let your interest in new-energy politics end on Election Day. Write to your representatives, urging that less money be spent on secret military research and more on new-energy research. Point out that this makes sense in an era of budget deficits and government cutbacks.

• Write letters to the news editor of your local or regional newspaper, and to a local radio or television station, asking for fair coverage of new-energy developments.

• If you are building or renovating a home—or if you know someone who is—explore new-energy options as they become available. You might find that even if the initial cost is greater than that of standard energy hardware, the energy savings will allow the device to pay for itself.

Free energy. Freedom from slavery to the narrow worldview of materialistic science. Freedom from the deeply grooved path of outmoded thinking. Freedom to find a way of tapping into that background energy out of which everything is created. Researcher Hal Puthoff compels us to realize our role in shaping reality: "Only the future will reveal to what use humanity will eventually put this remaining fire of the gods."

Will it happen? We decide.

Resource List

The search for new sources of decentralized, nonpolluting, readily available energy has brought together individuals from all over the globe. This common search has led these individuals to publish books and magazines, form organizations, and set up computer bulletin boards, the better to exchange ideas and find support.

To encourage you to learn more about the coming energy revolution, I have prepared the following resource list. The bookshelf lists specific books that I think would provide a good background, including addresses for books that might be hard to find. The newsstand lists publishers and lenders of books, magazines, and tapes. Remember, knowledge is power—the power to change our lives.

THE NEW-ENERGY BOOKSHELF

Alexandersson, Olof, *Living Water: Viktor Schauberger and the Secrets of Natural Energy*, Turnstone Press Ltd., Wellingborough, England, 1990. Distributed in North America by Borderland Sciences Research Foundation (see the newsstand section).

Aspden, Harold, *Physics Unified*, Sabberton Publications, Southampton, England, 1980 (see the newsstand section).

Baumgartner, W., and Dale Pond, *Tele-Geo-Dynamics*, Delta Spectrum Research, Valentine NE, 1993 (see the newsstand section).

Bearden, Thomas, *Excalibur Briefing*, Strawberry Hill Press, San Francisco, 1980.

Bearden, Thomas, *Gravitobiology*, Tesla Book Company, Chula Vista CA, 1991 (see the newsstand section).

Bearden, Thomas, *The New Tesla Electromagnetics and the Secrets of Electrical Free Energy*, Tesla Book Company, Chula Vista CA, 1990 (see the newsstand section).

Bearden, Thomas, and Michrowski, Andrew, *The Emerging Energy Science: An Annotated Bibliography*, Planetary Association for Clean Energy, Ottawa Ont., 1985 (see the newsstand section).

Becker, Robert O., *Cross Currents*, J.P. Tarcher Inc., Los Angeles, 1990.

Bedini, John C., *Bedini's Free Energy Generator*, Tesla Book Company, Chula Vista CA, 1984 (see the newsstand section).

Billings, Roger E., *The Hydrogen World View*, International Academy of Science, Independence MO, 1991 (see the newsstand section).

Boadella, David, *Wilhelm Reich: The Evolution of His Work*, Arkana, London, 1985.

Bockris, John O'Malley, *Energy: The Solar-Hydrogen Alternative*, John Wiley & Sons, Inc., New York, 1975.

Bockris, John O'Malley, T. Nejat Veziroglu, and Debbi Smith, *Solar Hydrogen Energy: The Power to Save the Earth*, Macdonald & Co., London, 1991.

Brown, Tom, ed., *The Hendershot Motor Mystery*, Borderland Sciences Research Foundation, Bayside CA, 1988 (see the newsstand section).

Cheney, Margaret, *Tesla: Man Out of Time*, Dell Publishing, New York, 1981.

Childress, David Hatcher, *Anti-Gravity and the World Grid*, Adventures Unlimited Press, Kempton IL, 1987 (see the newsstand section).

Childress, David Hatcher, *The Free Energy Device Handbook*, Adventures Unlimited Press, Kempton IL, 1994 (see the newsstand section).

Conti, Biagio, *Exotic Patents*, Conti Associates, Box 1014, Carmel NY, 10512, 1994.

Davidson, Dan A., *Energy: Breakthroughs to New Free Energy Devices*, RIVAS, Sierra Vista AZ, 1990 (see the newsstand section).

Davidson, Dan A., *Energy: Free Energy, the Aether and Electrification*, RIVAS, Sierra Vista AZ, 1992 (see the newsstand section).

Davidson, Dan A., *The Theta Device and Other Free Energy Patents*, RIVAS, Sierra Vista AZ, 1990 (see the newsstand section).

Davidson, John, *The Secret of the Creative Vacuum*, C.W. Daniel Co. Ltd., Essex, England, 1989.

Davidson, John, *Subtle Energy*, C.W. Daniel Co. Ltd., Essex, England, 1987.

Eisen, Jonathan, editor, *Suppressed Inventions and Other Discoveries*, Auckland Institute of Technology Press, Auckland, New Zealand, 1994.

Ford, R.A., *Homemade Lightning: Creative Experiments in Electricity*, Tab Books, Blue Ridge Summit PA, 1991.

Fox, Hal, *Cold Fusion Impact in the Enhanced Energy Age*, Fusion Information Center, Salt Lake City UT, 1992 (see the newsstand section). Bibliography on computer disk available in English, Russian, and Spanish.

Freeman, John, *Suppressed and Incredible Inventions*, reprinted by Health Research, P.O. Box 70, Mokelumne Hill CA, 95245, 1987.

Graneau, Peter, and Neal Graneau, *Newton Versus Einstein: How Matter Interacts With Matter*, Carlton Press, New York, 1993.

Hayes, Jeffrey A., *Boundary-Layer Breakthrough: The Bladeless Tesla Turbine*, available through Twenty First Century Books, Breckenridge CO, 1990 (see the newsstand section).

Hayes, Jeffrey A., *Tesla's Engine: A New Dimension for Power*, Tesla Engine Builders' Association, Milwaukee WI, 1994 (see the newsstand section).

Johnson, Ben, editor, *My Inventions: The Autobiography of Nikola Tesla*, Hart Brothers, Williston VT, 1982.

Kelly, Don, *The Manual of Free Energy Devices and Systems*, Cadake Industries Inc., Clayton GA, 1987 (see the newsstand section).

King, Moray B., *Tapping the Zero-Point Energy*, Paraclete Publishing, P.O. Box 859, Provo UT, 84603, 1989.

King, Serge K., *Earth Energies: A Quest for the Hidden Power of the Planet*, Quest Books, Wheaton IL, 1992.

Kuhn, Thomas S., *The Structure of Scientific Revolutions*, University of Chicago Press, Chicago IL, 1970.

LaViolette, Paul, *Beyond the Big Bang*, Box 388, Rochester, VT 05767, 1995.

Lindemann, Peter A., *A History of Free Energy Discoveries*, Borderland Sciences Research Foundation, Bayside CA, 1986 (see the newsstand section).

Mallove, Eugene F., *Fire From Ice: Searching for the Truth Behind the Cold Fusion Furor*, John Wiley & Sons, Inc., New York, 1991.

Manning, Jeane, and Pierre Sinclaire, *The Granite Man and the Butterfly: The David Hamel Story*, Project Magnet Inc., Box 839, 9037 Royal Street, Fort Langley B.C. V1M 2S2, 1995.

Michrowski, Andrew, *New Energy Technology*, Planetary Association for Clean Energy, Ottawa Ont., 1988 (see the newsstand section).

Moore, Clara Bloomfield, *Keely and His Discoveries*, originally published 1893, available from Delta Spectrum Research Inc., Valentine NE (see the newsstand section).

Moray, T. Henry, and John Moray, *The Sea of Energy*, Cosray Research Institute, P.O. Box 651045, Salt Lake City UT 84165-1045, 1978.

Nieper, Hans A., *Dr. Nieper's Revolution in Technology, Medicine and Society*, MIT Verlag, Oldenburg, Germany, 1983.

O'Leary, Brian, *Miracle in the Void*, Kampapua'a Press, 1993 South Kihei Road, Suite 21-100, Kihei HI 96753, 1996.

O'Leary, Brian, *The Second Coming of Science*, North Atlantic Books, Berkeley CA, 1992.

O'Neill, John J., *Prodigal Genius: The Life of Nikola Tesla*, Angriff Press, Hollywood CA, 1978.

Pond, Dale, editor, *Universal Laws Never Before Revealed: Keely's Secrets*, The Message Company, Santa Fe NM, 1995.

Resines, Jorge, *Some Free Energy Devices*, Borderland Sciences Research Foundation, Bayside CA, 1989 (see the newsstand section).

Russell, Walter, and Lao Russell, *Atomic Suicide*, University of Science and Philosophy, Waynesboro VA, 1957 (see the newsstand section).

Schaffranke, Rolf, *Ether-Technology*, Cadake Industries Inc., Clayton GA, 1977 (see the newsstand section).

Seike, Shinichi, *The Principles of Ultra-Relativity*, Gravity Research Lab, P.O. Box 33, Uwajima City, Ehime (798) Japan, 1978.

Tewari, Paramahamsa, *Beyond Matter*, Printwell Publications, Lekh Raj Nagar, Aligarh-202001, India, 1984.

Valone, Thomas, *Electrogravitics Systems*, Integrity Research Institute, Washington DC, 1994 (see the newsstand section).

Valone, Thomas, *The One-Piece Faraday Generator: Theory and Experiment*, Integrity Research Institute, Washington DC, 1987 (see the newsstand section).

Winter, Dan, and others, *Alphabet of the Heart: Sacred Geometry*, San Graal School, Waynesville NC, 1992 (see the newsstand section).

Wiseman, George, *The Energy Conserver Method*, Eagle Research, Yahk B.C., 1994 (see the newsstand section).

THE NEW-ENERGY NEWSSTAND

Adventures Unlimited Press
P.O. Box 74
303 Main Street
Kempton IL 60946
Phone: (815) 253-6390
Fax: (815) 253-6300

This publisher explores new-energy topics, from antigravity to Tesla technology.

The A/G Society
c/o George Overton
2 Thames View
Kelmcott GL7 3AG
England

This is a small network of antigravity and new-energy researchers.

All Source Digest
c/o Byron Peck
Box 596
Morton WA 98356

This alternate technology newsletter is published sporadically.

Alternative Energy Network
119 South Fairfax Street
Alexandria VA 22314
Phone: (707) 683-0774

This network provides daily news summaries on alternative fuels and transportation, as well as news on clean air, ozone, and global warming.

ASPS (Associazione Sviluppo Propulsione Spaziale)
c/o Dr. Emidio Laureti
Dipartimento RA-1
Via N. Martoglio 22
00137 Roma Italia
Phone/fax: 0039-6-87131068

This group works to develop private-sector, new-energy space exploration.

As You Like It Library
915 East Pine # 401
Seattle WA 98722
Phone: (206) 329-1794

This privately funded library, open to public memberships, has a section on new-energy technology.

Backcountry Productions
831 Alpine Street
Longmont CO 80501
Phone: (303) 772-8358

This company provides videotapes and audiotapes of new-energy conferences held in Denver in 1993 and 1994.

Ballard Power Systems
980 West First Street
North Vancouver B.C. V7P 3N4
Phone: (604) 986-9367
Fax: (604) 986-3252

This publicly traded company is developing fuel-cell systems for large power plants and transportation systems.

Borderland Sciences Research Foundation
P.O. Box 220
Bayside CA 95524
Phone: (707) 825-7733
Fax: (707) 825-7779
E-mail: bsrf @ asis.com
Internet: http://www.asis.com

*This organization publishes **Borderlands** magazine, a quarterly that has served people interested in nonconventional research, including new-energy research, since 1945.*

Brewer International Science Library
325 North Central Avenue
Richland Center WI 53581
Phone: (608) 647-6513
Fax: (608) 647-6797

This privately funded library contains a new-energy section.

Cadake Industries, Inc.
P.O. Box 1490
Clayton GA 30525
Phone: (706) 782-7714

This publisher handles unusual titles, in-cluding new-energy topics.

Center for Frontier Sciences
Temple University
Ritter Hall 003-00
Philadelphia PA 19122
Phone: (215) 204-8487
Fax: (215) 204-5553

This organization is a bridge between orthodox and unorthodox scientists. It sponsors roundtables for scientists and publishes a semiannual newsletter, Fron-tier Perspectives.

Cheyenne Mountain BBS
c/o Warren York
7101 North Mesa, Suite 133
El Paso TX 79912
Phone: (915) 585-3674
Internet: @ primenet.com

This computer bulletin board contains new-energy information useful to novices and veterans alike.

Citizens' Energy Council
P.O. Box U
Hewitt NJ 07421
Phone: (201) 728-7835
Fax: (201) 728-7664

This information clearinghouse publishes a newsletter, The Messenger, at irregu-lar intervals.

Cold Fusion Newsletter
c/o Wayne Green
70 Route 202 North
Peterborough NH 03458
Phone: (603) 525-4747
          (800) 677-8838 to subscribe
Fax: (603) 924-8613

This newsletter is published on a month-ly or near-monthly basis.

Contact Network International
P.O. Box 66
8400 AB Gorredijk
Netherlands
Phone: 31 5133 5567

This is a resource base of print and elec-tronic media on clean energy systems in the Netherlands, with an educational focus.

Cosmic Energy Association of Japan
c/o Dr. Masayoshi Ihara
37-2 Nisigoshonouti, Kinugasa
Kitaku, Kyoto
Japan

This association does new-energy research.

Cosmic Energy Foundation (De Stichting Kosmiese Energie)
c/o Martin Holwerda
Neptunuslaan 11
3318 El Dordrecht
Netherlands
Phone: (31) 078-170405

This group promotes the practical develop-ment of space energy. Program proceed-ings since 1987 are available in English.

Danish Institute for Ecological Techniques
c/o Margrete Schou
Lyngbyvej 424, 2.th.
DK-2820 Gentofte
Denmark
Phone: (45) 4 2892049
Fax: (45) 2891865

This group does new-energy research and publishes a journal, Difot-nyt.

Delta Spectrum Research Inc.
c/o Dale Pond
P.O. Box 316
Valentine NE 69201

Pond, a veteran Keely researcher, writes books and has put the bulk of Keely's work onto computer disks.

Eagle Research
P.O. Box 10
Yahk B.C. V0B 2P0
Phone: (604) 424-5488

US address:
P.O. Box 145
Eastport ID 83826

This is George Wiseman's research group. It sells how-to publications on alternate energy and fuel-saving devices.

Electric Spacecraft Journal
73 Sunlight Drive
Leicester NC 28748
Phone: (704) 683-1280
Fax: (704) 683-3511

This journal covers how-to material for experimenters, and advanced scientific and technical concepts.

Electrifying Times
63600 Deschutes Road
Bend OR 97701
Phone: (503) 388-1908
Fax: (503) 382-0384

This journal covers solar and electric car events, battery technologies, and other new-energy material.

EXPLORE MORE
c/o Chrystyne Jackson
P.O. Box 1508
Mt. Vernon WA 98273

This bimonthly magazine covers some new-energy topics.

FUNDPAC (Fundación para el Avance del Conociemento)
Allayme 1719
San José, Guaymallén, Argentina
Phone: 54 61 242 770

This is the Latin American arm of the Ottawa-based Planetary Association for Clean Energy (see entry).

Fusion Information Center
P.O. Box 58639
Salt Lake City UT 84158-8639
Phone: (801) 583-6232
Fax: (801) 583-6245

Hal Fox edits this group's monthly cold-fusion newsletter, Fusion Facts.

Future Technology Intelligence Report
c/o Dr. Antony C. Sutton

P.O. Box 423652
San Francisco CA 94142-3652

This newsletter—a bit expensive at $250 a year—reports on new energy and other tehcnologies.

German Association of Vacuum Field Energy
c/o Klaus Van Dollen
Nelkenstra B Str. 50
D-26121 Oldenburg
Germany
Fax: 011 49 441 980 7777

This group has deposited educational/scientific materials at the Brewer International Science Library (see entry) and at:

Baylor University
Glasscock Energy Research Center
Attn: Dr. Dudley J. Burton
B.U. Box 7266
Waco TX 76798-7266
Phone: (817) 755-3405
Fax: (817) 755-3878

Global Sciences
c/o Dean Stonier
3273 East 119th Place
Thornton CO 80233
Phone: (303) 452-9300
Fax: (303) 457-8269

This group holds two conferences a year, which include an occasional new-energy speaker.

H & A Industries
Route 2, Box E-35
Bowling Green MO 63334

This company sells manuals on unorthodox energy and automotive research.

Mark M. Hendershot
16541 Redmond Way # 160
Redmond WA 98052

Hendershot sells a packet of information and photos about the device invented by his father, Lester.

The Holo Tec Clean Energy Technology
c/o André Waser
Wesemlinrain # 12
6006 Luzern
Switzerland
Phone: 41-41 360 4485
Fax: 41-41 360 4486

This bookstore disseminates information on new energy.

Home Power
P.O. Box 520
Ashland OR 97520
Phone: (916) 475-3179

This bimonthly magazine is aimed at solar- and wind-power do-it-yourselfers, but also covers a broad range of energy issues. It contains a column, "The Wizard Speaks," that invites inventors to send in new-energy device prototypes for testing.

Horizon Technology
c/o Gary Hawkins
2442 NW Market Street # 274
Seattle WA 98107

This group sends out technology updates periodically on a wide range of topics.

Infinite Energy
c/o Eugene Mallove
P.O. Box 2816
Concord NH 03302-2816
Phone: (603) 228-4516
Fax: (603) 224-5975
Internet:
76570.2270@compuserve.com

This readable, informative magazine covers cold fusion and other new-energy developments.

Institute for Fundamental Physics
East-West Publishers
Morellenfeldgasse 16
A8010 Graz
Austria
Fax: (43) 0316\835015

This institute, Stefan Marinov's one-man

*research lab, publishes **Deutsche Physik**, which contains some English-language writeups of various experiments.*

Institute for New Energy
P.O. Box 58639
Salt Lake City UT 84158-8639
Phone: (801) 583-6232
Fax: (801) 583-2963
Internet: www.padrak.com/INE/

The institute publishes New Energy News, a monthly magazine edited by Hal Fox. It is a best-buy for up-to-date international new-energy information.

Integrity Research Institute
c/o Thomas Valone
1413 K Street NW, Suite 204
Washington DC 20005
Phone: (202) 452-7674
    (800) 329-8416

This institute does research on energy and on the biological effects of energetic radiation, such as the electromagnetic fields that emanate from computers.

International Academy of Science
TechCenter, Suite 1000
26900 East Pink Hill Road
Independence MO 64057-3284

Roger Billings works on hydrogen technology at this research center.

International Association of Hydrogen Energy
P.O. Box 248266
Coral Gables FL 33124

This association publishes the International Journal of Hydrogen Energy, which can be ordered through:

Elsevier Science
660 White Plains Road
Tarrytown NY 10591-5153

International Association for New Science
1304 South College Avenue
Fort Collins CO 80524
Phone: (970) 482-3731
Fax: (970) 482-3130

This group hosts conferences on new energy and other topics.

International Science Forum Network
170 West 74th Street # 904
New York NY 10023
Fax: (212) 595-5846

This network provides audiotapes of new-energy meetings.

International Tesla Society
P.O. Box 5636
Colorado Springs CO 80931
Phone: (719) 475-0918
Fax: (719) 475-0582

*This society issues a catalog of books and tapes. It also sponsors annual symposia, runs a museum, and does educational outreach. The quarterly magazine **Extraordinary Science** is included in the membership fee.*

Keelynet BBS
c/o Jerry Decker, Vanguard Sciences
P.O. Box 1031
Mesquite TX 75150
Phone: (214) 324-3501
Internet:
www.protree.com/keelynet/

This computer bulletin board provides information on the work of John Keely and other, more recent, new-energy researchers.

Newtext: The Flashpoints
c/o Dr. Nick Begich
P.O. Box 201393
Anchorage AK 99520
Phone: (907) 249-9111

This bimonthly newsletter contains hard-to-find information about new energy and other topics.

Nexus
P.O. Box 30
Mapleton, Queensland 4560
Australia
Phone: 61 (0) 7 5442 9280

Fax: 61 (0) 7 5442 9381
E-mail: nexus at peg.apc.org

US address:
P.O. Box 177
Kempton IL 60946-0177
Phone: (815) 253-6464
Fax: (815) 253-6300
Internet:
http://www.peq.apc.org~nexus/

This magazine covers new energy, as well as other suppressed technologies.

Norweigan Institute for Energy Technology Assessment
c/o Jorn Erik Ommang
Nordeidevein 39
5060 Soreidegrend-Bergen
Norway
Phone: 47 (55) 124-718
Fax: 47 (55) 998-474

This institute does new-energy research.

Orgone Biophysical Research Lab
P.O. Box 1148
Ashland, OR 97520
Phone: (541) 552-0118

*This organization does experiments related to the work of Wilhelm Reich and publishes a journal, **Pulse of the Planet**.*

People's Network Inc.
3 River Street
White Springs FL 32096
Phone: (904) 543-5648

*This group offers a book catalog for a stamped, self-addressed envelope, and publishes a quarterly, **For the People**. It also sponsors a three-hour talk show on worldwide shortwave radio from 2 to 5 P.M. Eastern time at 9.495 and from 10 P.M. to midnight at 5.760.*

Planetartsche Vereinigung für Saubere Energie, Inc.
Feyermühler Strasse 12
D-53894 Mechernich
Federal Republic of Germany
Phone: (49) 2443 2682

Fax: (49) 2443 8221
E-mail: 100276.261@compuserv.com

This is the European arm of the Ottawa-based Planetary Association for Clean Energy (see entry).

Planetary Association for Clean Energy (PACE)
100 Bronson Avenue, Suite 1001
Ottawa Ont. K1R 6G8
Phone: (613) 236-6265
Fax (613) 235-5876

This not-for-profit corporation was founded in 1979 to encourage the development of clean energy systems. The PACE newsletter is published irregularly. PACE also has a new-energy computer database.

Quantum Realities
c/o Warren York
7101 North Mesa, Suite 133
El Paso TX 79912
Infonet: @ primenet.com

This newsletter covers such topics as new energy and subatomic physics.

Radio Free America

This talk show, which often deals with new-energy topics, is heard internationally from 10 P.M. to 12 midnight Eastern time on WWCR 5810 kilohertz short-wave.

Real Goods Trading Corporation
966 Mazzoni Street
Ukiah CA 95482

This company has a catalog of alternative energy products for home use. It is an excellent resource for the transition time before cheaper, super-efficient new-energy products go on the market.

Rex Research
Box 19250
Jean NV 89019

Rex has been collecting new-energy information for decades, and will sell you photocopies of documents and articles. A catalog is available.

RIVAS (Realistic Inspired Vital Appropriate Solutions)
P.O. Box 1090
Sierra Vista AZ 85636

Send a stamped, self-addressed envelope for a catalog of Dan Davidson's books and papers—clear writing and ongoing updates from an electrical engineer and nuclear physicist who is a long-time new-energy researcher.

Rocky Mountain Institute
c/o Dr. Amory Lovins
1739 Snowmass Creek Road
Snowmass CO 81654-9199
Phone: (970) 927-3851
Fax: (970) 927-4178

This organization publishes a newsletter on energy efficiency news.

Russian Academy of Science
P.O. Box 169 Erzion Centre
105077 Moscow, Russia

This academy organizes cold-fusion conferences.

Sabberton Publications
P.O. Box 35
Southampton SSO9 7BU
England

This publisher sells a series of Energy Science reports and books by Harold Aspden.

San Graal School for Sacred Geometry
c/o Daniel Winter
137 Biodome Drive
Waynesville NC 28786
Fax: (704) 926-9041
E-mail: DanWinter@aol.com

This group sells books and videos about the more esoteric aspects of new energy.

Scandinavian Association of Vacuum Field Energy
Gräbrödersgaten 10
S-211 21 Malmö
Sweden
Fax: 46-40 117898

This association organizes meetings on new energy.

Scientific and Medical Network
c/o David Lorimer
Lesser Halings, Tile House Lane
Denham-Oxbridge UB9 5DG
England
E-mail: Compu.serve 100114, 1637 at smnet.demon.co.uk

This research group does work on new energy and other topics.

Space Energy Association
c/o Jim Kettner
P.O. Box 11436
Clearwater FL 34617
Phone: (813) 442-3923
Fax: (813) 446-5290

*This group publishes the quarterly **Space Energy Journal**. It provides top-quality new-energy information and construction plans.*

Space Research Institute
c/o Shinichi Seike
Box 33
Uwajima, Ehime 79
Japan

Seike is a pioneer in the new-energy field.

The Stan Deyo Newsletter
P.O. Box 71
Kalamunda
Western Australia 6076

This monthly newsletter covers a wide range of topics, including anitgravity and new energy. Deyo is a long-time researcher in these fields.

S.T.E.V.E.N. Foundation (Solar Technology and Energy for Vital Economic Needs)
414 Triphammer Road
Ithaca NY 14850
Phone: (607) 257-7109

Through this foundation, Jaroslav Vanek develops inexpensive solar, wind, and hy-
draulic technologies that can be built by and for poor people. A videotape is available for a donation of US $20. Contruction plans are also available.*

Swiss Association for Free Energy
c/o Werner Rusterholz
P.O. Box 10
5704 Egliswill
Switzerland
Phone: (41) 064 55 10 83

The proceedings of this group's 1989 conference are available in English.

Tesla Book Company
P.O. Box 121873
Chula Vista CA 91912
Phone: (619) 585-8481
           (800) 398-2056

This publisher has both videotapes and an extensive book list. The catalog is free.

Tesla Coil Builders' Association
3 Amy Lane
Queensbury NY 12804
Phone: (518) 792-1003

This educational organization, founded in 1982, provides information on Tesla coils—high voltage/high frequency transformers. It publishes a quarterly newsletter.

Tesla Engine Builders' Association
5464 North Port Washington Road
Suite 293
Milwaukee WI 53217

This is a not-for-profit information clearinghouse for Tesla turbine builders. Send a self-addressed, stamped envelope for more information.

Tesla HAMNet
c/o Mike DiPersio (KC2Q)
P.O. Box 357
Bradley Beach NJ 07720

This is a network of ham radio operators who are interested in Nikola Tesla. It is at 14.297 Mhz.

Tesla Memorial Society
c/o William H. Terbo
453 Martin Road
Lackawanna NY 14218

This group is attempting to have a Nikola Tesla display placed in the Smithsonian (see Chapter 2). Its directors include authors and scientists. Its publications include The Tesla Journal.

Timeless Voyager Radio
P.O. Box 6678
Santa Barbara CA 93160
Phone: (805) 964-3301
(800) 576-TIME (catalog orders only)
Fax: (805) 683-4456

New energy is one part of this radio service's catalog of program episodes, books, and audiotapes.

Twenty First Century Books
c/o Gary L. Peterson
P.O. Box 2001
Breckenridge CO 80424-2001
Phone: (970) 453-9293

This "Tesla technology resource" can get hard-to-find publications. A catalog is available.

21st Century Science and Technology
P.O. Box 16285
Washington DC 20041

This magazine provides information on advanced technologies and science policy.

Ukrainian Association of Vacuum Field Energy
Kyiv Institute of Energy Saving
  Problems
11 Pokrovskaya Street
254070 Kyiv
Ukraine
Phone: 7 44 417-0412

This organization promotes research and the spread of information on new energy.

United States Psychotronics Association
P.O. Box 354

Wilmette IL 60091
Phone: (708) 733-0116
Fax: (708) 733-0117

This association deals with a broad scope of topics, which often includes a new-energy segment. Its audio-visual archives include lectures by such speakers as Thomas Bearden and Ed Skilling.

University of Science and Philosophy
Swannanoa
P.O. Box 520
Waynesboro VA 22980
Phone: (800) 882-LOVE (orders only)

This university-without-walls offers seminars, a correspondence course, and the newsletter Light Waves on the teachings of Walter and Lao Russell.

Untapped Technology in Review
The UTR Press
P.O. Box 5185
Mesa AZ 85211

This is an excellent resource for serious researchers and laypeople alike. It summarizes and references other publications.

Visions Unlimited
c/o John Thomas
373 Rock Beach Road
Rochester NY 14617

This company offers publications on the work of long-time inventor John Searl of England.

Wholistic Research Company
Bright Haven
Robin's Lane, Lolworth
Cambridge CB3 8HH
England
Phone: 44 (1954) 781074

This company sells the New Science series of books.

Wireless Engineering
c/o Toby Grotz
E-mail: wireless@rmii.com
Internet: www.yampa.com//
wireless/

This is a design and management service for new-energy inventors, but the novice researcher can find useful information through the Internet site.

Glossary

Italicized words are defined elsewhere in the Glossary.

Aether. The background substance of the universe, now thought to be in a constant spiralling motion, that of a *vortex*. It is the basic substance out of which the universe is made, and it gives rise to *space energy*.

Alternating current (AC). Electricity that flows back and forth in a regular rhythm. In the United States and Canada, standard household current changes direction sixty times a second.

Antigravity. A force that opposes or cancels gravity, which is the force that pulls or pushes all objects on the earth's surface towards the planet's center. Under certain conditions, *space energy* can produce an antigravity effect.

Casimir effect. The tendency for two perfectly smooth metal surfaces placed very near each other to come closer together. It is thought that *space energy* causes this effect.

Cavitation. The formation of cavities or bubbles in liquids, and the collapse of those bubbles. Shock waves are created when the bubbles that form in a low-pressure section of a liquid-carrying pipe collapse upon being carried to a high-pressure section. Cavitation is also called water hammer.

Cermet. A composite made of ceramic and metal that is used in some *space energy* research.

Charge cluster. A ring-shaped structure made of tightly packed electrons.

Chemical energy. Energy produced by burning, such as that produced by oil or coal.

Closed system. A system in which a finite amount of energy is available. An *internal combustion engine* can be said to operate in a closed system.

Cold fusion. The joining together of atomic nuclei under room-temperature conditions in order to release energy.

Direct current (DC). Electricity that flows in one direction.

Dynamic electricity. Electricity in motion, such as the flow of current through a wire.

Earth resonance. The sending of electric pulses through the earth at the same rate at which the earth itself vibrates in order to build up large waves of energy. Such energy could be picked up at a distance by an antenna.

Electrolysis. The breakdown of water into its component parts, oxygen and *hydrogen*, by passing an electric current through it.

Electrolytic cell. A cell that contains an electricity-conducting liquid in which two metal wires or plates are suspended. *Cold fusion* takes place in such a cell.

Electromagnetics. The physics of electricity and magnetism.

Entropy. The idea that matter and energy are always becoming more and more disorganized. It is the opposite of *negentropy*.

Fission. The splitting of an atom's nucleus in order to release energy.

Fossil fuel. Fuel produced by the decay of prehistoric plants and animals deep underground. Oil, coal, and natural gas are all fossil fuels.

Free energy. See *space energy*.

Fuel cell. A cell in which a gas, such as *hydrogen*, is used to create electricity.

Generator. A device that converts mechanical energy into electricity.

Heat pump. A device that heats a structure by drawing heat from the surrounding soil, air, or water.

Heat technology. Devices that derive energy from a difference in temperature, for example, by changing a liquid into a gas. *New-energy* heat technology uses liquids that change into gases at low temperatures.

Hot fusion. The joining together of atomic nuclei under conditions of high heat and pressure in order to release energy.

Hydrogen. The lightest known element, consisting of one proton and one electron. Its abundance and wide distribution over the earth's surface means that hydrogen can be used to provide inexpensive, decentralized energy.

Hydropower. The use of moving water to generate electricity. Standard hydropower uses environmentally destructive dams, but *new-energy* hydropower uses devices that do not harm the environment.

Implosion generator. A *generator* that uses an inward-spiralling *vortex* movement to produce power.

Induction. The electrification of a wire that occurs when the wire is placed near a moving magnetic field.

Internal combustion engine. Vehicle engines in which refined *fossil fuel,* such as gasoline, is burned. The energy released by this burning is transformed into a rotary motion that propels the vehicle's wheels.

Magnet device. A device that uses magnets to turn *space energy* into electricity.

Magnetic drag. A problem in standard *generators* in which residual magnetism slows down the rotor, which is the part that either moves the magnets past the coils of wire or the wire coils past the magnets—see *induction*. This drag reduces the generator's output.

Metal hydride. A combination of metals that allows for the safe storage of *hydrogen*. Under proper conditions, the hydride soaks up hydrogen and holds it until the hydrogen is released on demand.

Motor. A device that converts electricity or *chemical energy* into mechanical energy.

Negentropy. The idea that matter and energy can organize themselves. It is the opposite of *entropy*.

New energy. Energy that comes from nonconventional sources, preferably that which can be produced on a decentralized basis. *Cold fusion, heat technology, hydrogen* technology, low-impact *hydropower*, and *space energy* are all examples of new energy.

Nuclear energy. Energy produced by either the breaking apart or bringing together of atomic nuclei. Also see *cold fusion, fission,* and *hot fusion*.

Open system. A system in which an infinite amount of energy is available. A *space energy* device can be said to operate in an open system.

Overunity. A condition in which there is more energy going out of a device than coming in.

Perpetual-motion machine. A device that, once set in motion, continues to operate without an outside source of energy within a *closed system*. Such a device is impossible to build.

Quantum mechanics. The branch of science that deals with protons, electrons, and other basic particles of matter.

Self-oscillation. The continuous shaking or vibration of a magnetic field.

Solid-state devices. Devices that contain no moving parts.

Sonoluminescence. The light that is given off when ultrasound waves are pumped into tiny bubbles in a liquid, and the bubbles collapse violently.

Space energy. Energy that consists of electrical fluctuations in the *aether*. It is present everywhere in the universe, including the earth, but can only be put to use through the use of specially designed devices.

Static electricity. Electricity at rest, such as the electric charge that builds up on a plastic comb.

Superconductivity. The state of a wire when it suddenly loses resistance, the force that keeps current from passing through the wire. Superconductivity normally occurs only at very low temperatures, but it could help create a practical source of electrical power if it could be made to occur at room temperature.

Turbine. A machine that uses a stream of either gas or liquid to turn a shaft, such as the rotor of a *generator*.

Vortex. A three-dimensional spiral that creates a funnel of energy, such as a tornado. Motion along such a spiral can be outward, in which energy is dissipated, or inward, in which energy is created.

Zero-point energy. See *space energy*.

Permissions

The photograph on page 25 is used courtesy of John W. Wagner.

The photographs on pages 52, 57, 58, 65, 84, 90, 113, 123, 138, and 144 are used courtesy of Jeane Manning.

The photographs on pages 63, 73, 89, and 125 are used courtesy of Toby Grotz, Institute for New Energy.

The photograph on page 132 is used courtesy of Nova Energy.

The quotation on page 12 is from *Newsletter of the Planetary Association for Clean Energy*. May 1980, Vol. 2 No. 3 p. 19. Reprinted by permission of the Planetary Association for Clean Energy.

The quotation on page 14 is from *Space Energy Journal*. December 1994, Vol. 5 No. 4 p. 35. Reprinted by permission of the Space Energy Association.

The quotation on pages 43–44 is from *Energy: Breakthroughs to New Free Energy Technologies*. RIVAS, 1990 p. 2. Reprinted by permission of Dan A. Davidson.

The quotation on page 75 is from *Space Energy Newsletter*. March 1993 Vol. IV No. 1 p. 1. Reprinted by permission of the Space Energy Association.

The quotation on page 80 is from *Proceedings of the 1991 Intersociety Energy Conversion Engineering Conference*. Vol. 4 Pt. 1 p. 374. Reprinted by permission of the American Nuclear Society. Copyright 1991 by the American Nuclear Society, Inc., LaGrange Park, Illinois.

The quotation on pages 158–159 is from *Proceedings of the Second International Symposium on Nonconventional Energy Technology*. 1983 p. 143. Reprinted by permission of the Planetary Association for Clean Energy.

The quotation on page 162 is from *Proceedings of the Second International Symposium on Nonconventional Energy Technology*. 1983 p. 144. Reprinted by permission of the Planetary Association for Clean Energy.

The quotation on page 183 is from *The Secret of the Creative Vacuum*. 1989 p. 243. Reprinted by permission of C. W. Daniel Company Ltd.

Bibliography

Preface

Chowdhuri, P., T.W. Linton, and J.A. Phillips. "A Rotating Flux Compressor for Energy Conversion." *Space Energy Journal*, special issue for the International Symposium on Energy, 12 May 1994, 3.

Davidson, Dan A. *Energy: Breakthroughs to New Free Energy Devices.* Sierra Vista, AZ: RIVAS, 1990.

Fox, Hal. "Comets and NEN Plans for Dinosaur Thinking." *New Energy News*, August 1994, 3–5.

Peterson, John L. *The Road to 2012: Looking Toward the Next Two Decades.* Arlington, VA: Arlington Institute, 1992.

Sabar, Ariel. "Greenhouse 101." *Whole Earth Review*, Winter 1994, 16.

Storms, Edmund. "Cold Fusion Alive, Growing; So Why Is LANL Ignoring It?" *Los Alamos Monitor*, 3 April 1994.

Chapter 1   *Quantum Leap*

Bird, Christopher. "Culture Control—in the Hands of Time." In *Suppressed Inventions and Other Discoveries*, ed. Jonathan Eisen, 31–33. Auckland,

New Zealand: Auckland Institute of Technology Press, 1994.

Borson, Daniel, and others. "A Decade of Decline: The Degeneration of Nuclear Power in the 1980s and the Emergence of Safer Energy Alternatives." Reprint of chapter from *Public Citizen, Critical Mass Energy Project*, 404–406. Washington, DC: Public Citizen's Critical Mass Energy Project, 1989.

Boyer, Timothy. "The Classical Vacuum." *Scientific American* Vol. 253 No. 2 (August 1985): 70–79.

Brower, Michael. *Cool Energy.* Cambridge, MA: MIT Press, 1992.

The Campaign for Nuclear Phaseout. *Financial Meltdown: Government Subsidies for the Nuclear Industry.* Ottawa, Ont.: The Campaign for Nuclear Phaseout, 1993.

"Chained to Reactors." *The Economist* Vol. 318 No. 7692 (2 February 1991): 59–60.

Chukanov, K. "Energy Source of the 21st Century." Salt Lake City, UT, 1993.

Cole, Daniel C., and Harold Puthoff. "Extracting Energy and Heat from the Vacuum." *Physical Review E* Vol. 48 No. 2 (August 1993): 1562–1565.

Edwards, Gordon. "Cost Disadvantages of Expanding the Nuclear Power Industry." *The Canadian Business Review* Vol. 9 No. 1 (Spring 1982): 19–30.

Emsley, John. "Energy and Fuels." *New Scientist* Vol. 141 No. 1908 (15 January 1994): 1–4.

The Europa World Yearbook. Vol. 1. London: Europa Publications Ltd., 1995.

Flavin, Christopher, and Nicholas Lenssen. *Power Surge: Guide to the Coming Energy Revolution.* New York: W.W. Norton & Co., 1994.

Forward, R.L. "Extracting Electrical Energy from the Vacuum by Cohesion of Charged Foliated Conductors." *Physical Review B* Vol. 30 No. 4 (August 1984): 1700.

Fox, Hal. *Cold Fusion Impact in the Enhanced Energy Age.* Salt Lake City, UT: Fusion Information Center, 1992.

Fox, Hal. "Let's Declare War!" *New Energy News,* October 1994, 3.

Golob, Richard, and Eric Brus, eds. *The Almanac of Science and Technology.* Orlando, FL: Harcourt Brace Jovanovich, 1991.

Greenpeace, Ltd. *Questions and Answers on Nuclear Energy.* London, Ont.: 1989.

Grotz, Toby. "Finding the Energy of the Future." In *Proceedings of the International Symposium on New Energy,* in Denver, 12–15 May 1994. Fort Collins, CO: Rocky Mountain Research Institute, 1994.

Harter, Walter. *Coal—The Rock That Burns.* New York: Elseweir/Nelson Books, 1979.

Hasslberger, Josef. "A New Awareness." *raum & zeit* (now *Explore!*) Vol. 3 No. 1 (1991): 65–67.

"Japanese Overunity Motor Commentary by T.E. Bearden." *Space*

Energy Journal Vol. V No. 4 (December 1994): 35.

Jean-Romain, Frisch, ed. *World Energy Horizons.* Montreal: World Energy Conference, 1989.

King, Llewellyn. "On Solving the Great Problems of the World." Paper presented at the Intersociety Energy Conversion Engineering Conference, Boston, 1991.

Lambertson, Wingate. "Phaseout of the Fossil Fuel Industries." *Explore More!* Vol. 1 No. 8 (1994): 12–13.

Lindemann, Peter A. "Thermodynamics and Free Energy." *Borderlands* Vol. L No. 3 (Fall 1994): 6–10.

Maglich, Bogdan. "Energy Debates and the Working Scientist." *Planetary Association for Clean Energy* Vol. 2 No. 2 (1980): 19.

Mallove, Eugene. *Fire From Ice.* New York: John Wiley & Sons, Inc., 1991.

Michrowski, Andrew. "Vacuum Energy Developments: The Related Physics of Bioenergetic Phenomena." *Planetary Association for Clean Energy* Vol. 6 No. 4 (1993): 12–18.

Michrowski, Andrew, commenting on his speech "Vacuum Energy Developments," at the International Symposium on New Energy in Denver, 12–15 May 1994.

Moore, Curtis, and Alan Miller. *Green Gold: Japan, Germany, the United States and the Race for Environmental Technology.* Boston: Beacon Press, 1994.

Petersen, John L. *The Road to 2012: Looking Toward the Next Two Decades.* Arlington, VA: Arlington Institute, 1992.

Puthoff, Harold. "Ground State of Hydrogen as a Zero-point-fluctuation-determined State." *Physcial Review D* Vol. 35 No. 10 (May 1987): 3266–3269.

Puthoff, Harold. "Gravity as a Zero-Point Fluctuation Force." *Physical Review A* Vol. 39 No. 5 (1 March 1989): 2333–2342.

Puthoff, Harold. "Everything for Nothing." *New Scientist* Vol. 127 No. 1727 (28 July 1990): 52–55.

Puthoff, Harold. "Zero-Point Energy." *Fusion Facts* Vol. 3 No. 3 (September 1991): 1–2.

Puthoff, Harold. "Quantum Fluctuations of Empty Space: A New Rosetta Stone of Physics?" *Frontier Perspectives*, Vol. 2 No. 2 (Fall/Winter 1991): 19–23.

Shoulders, Kenneth R. U.S. Patent No. 5,018,180. 21 May 1991. *Energy Conversion Using High Charge Density*.

Slesser, Malcolm, ed. *Dictionary of Energy*. New York: Nichols Publishing, 1988.

Storms, Edmund. "Cold Fusion Alive, Growing; So Why Is LANL Ignoring It?" *Los Alamos Monitor*, 3 April 1994.

U.S. Department of Defense. Small Business Innovation Research Program. *Request for Proposals*, AF Section 86-77, subsection 6, 193. Washington, DC, 1986.

Uvarov, E.B., D.R. Chapman, and Alan Issacs. *The Penguin Dictionary of Science*. New York: Penguin Books, 1979.

Chapter 2   Nikola Tesla— The Father of Free Energy

Asimov, Isaac. *Asimov's Biographical Encyclopedia of Science and Technology*. Garden City, NY: Doubleday & Co. Inc., 1982: 560–561.

Bird, Christopher, and Oliver Nichelson. "Great Scientist, Forgotten Genius Nikola Tesla." *New Age*, 1977, 36–44.

Broad, William J. "Tesla, a Bizarre Genius, Regains Aura of Greatness." *The New York Times*, 28 August 1984.

Cheney, Margaret. *Tesla: Man Out of Time*. New York: Dell Publishing, 1981.

Cohan, George M. *Thomas A. Edison: Miracle Man*. New York: VosBurgh's Orchestration Service, 1929.

Congressional Record. 97st Cong., 1st sess., 1981, Vol. 127, pt. 62. Representative Henry J. Nowak speaking to commemorate the birth of Nikola Tesla.

Congressional Record. 101st Cong., 2nd sess., 1990, Vol. 136, pt. 86. Senator Carl Levin speaking to commemorate the birth of Nikola Tesla.

Eisenberg, Anne. "The Art of the Scientific Insult." *Scientific American* Vol. 270 No. 6 (June 1994): 116.

Finn, B.S., and others. *Edison: Lighting a Revolution*. Washington, DC: Smithsonian Institute, 1979.

Hall, Stephen S. "Tesla: a Scientific Saint, Wizard or Carnival Sideman?" *Smithsonian*, June 1986, 121–134.

Kovak, Ron. "The Power Wave 1899–1991." *Electric Spacecraft Journal* Vol. 1 No. 3 (1991): 6–17.

Marvin, Carolyn. *When Old Technologies Were New*. New York and Oxford: Oxford University Press, 1988.

Michrowski, Andrew. "Vacuum Energy Developments." In *Proceedings of the International Symposium on New Energy*, in Denver, 16–18 April 1993. Fort Collins, CO: International Association for New Science, 1993, 407–417.

Nichelson, Oliver. "Nikola Tesla's Later Energy Generation Designs." In *Proceedings of the 26th Intersociety Energy Conversion Engineering Conference*, in Boston, 4–9 August 1991.

LaGrange Park, IL: American Nuclear Society, 1991, Vol. 4 No. 3: 433–438.

"Nikola Tesla and the Development of Electric Power at Niagara Falls." In *The Tesla Journal*, 4–11. Lackawanna, NY, 1989/1990.

O'Neill, John J. *Prodigal Genius: The Life of Nikola Tesla*. Hollywood, CA: Angriff Press, 1978.

Pribic, Nikola R. "Nikola Tesla—A Yugoslav Perspective." In *The Tesla Journal*, 59–61. Lackawanna, NY, 1989/1990.

Tesla, Nikola. "The Problem of Increasing Human Energy." *The Century Illustrated Monthly Magazine*, June 1900, 210.

Tesla, Nikola. *My Inventions: The Autobiography of Nikola Tesla*. Williston, VT: Hart Brothers, 1982. Originally appeared in *Electrical Experimenter*, 1919.

"Tesla's Tower." *Electric Spacecraft Journal* Vol. 1 No. 2 (1991): 13–21.

"World System of Wireless Transmission of Energy." *Nexus* Vol. 2 No. 3 (August/September 1994): 45.

Wright, Charles. "The Great AC/DC War." In *Proceedings of the 1988 International Tesla Symposium*, in Colorado Springs, CO, 28–31 July 1988. Colorado Springs, CO: International Tesla Society, 1988, 5–12.

Chapter 3   *Other Innovators— In Harmony With Nature*

Alexandersson, Olof. *Living Water: Viktor Schauberger and the Secrets of Natural Energy*. Wellingborough, England: Turnstone Press Ltd., 1982.

Binder, Timothy A. "Walter Russell's Perspectives on Free Energy and the Russell Optical Dynamo Generator." In *Proceedings of the International Symposium on New Energy*, in Denver, 12–15 May 1994. Fort Collins, CO: Rocky Mountain Research Institute, 1994, 74–97.

Boadella, David. *Wilhelm Reich: The Evolution of His Work*. London: Arkana, 1985.

Brown, Tom, ed. *The Hendershot Motor Mystery*. Bayside, CA: Borderland Sciences Research Foundation, 1988.

Burridge, Gaston. "The So-called Hendershot Motor." *Round Robin* (now *Borderlands*) Vol. XI No. 6 (March/April 1956): 1–6.

Burridge, Gaston. "Alchemist 1956?" *Fate*, 16 September 1956, 16–22.

Davidson, Dan A. *Energy: Free Energy, the Aether and Electrification*. Sierra Vista, AZ: RIVAS, 1992.

Davidson, John. *The Secret of the Creative Vacuum*. Essex, England: C.W. Daniel Co. Ltd., 1989.

Eden, Jerome. *Orgone Energy*. Hicksville, NY: Exposition Press, 1972.

Hendershot, Mark M. "An Inside View of The Hendershot Motor Mystery." *Extraordinary Science*, October/December 1994, 5–10.

Holwerda, Martin. "Hendershot and Prentice Generators." *Electric Spacecraft Journal* Vol. 1 No. 7 (1992): 33–34.

Jackson, Chrystyne, and Jeane Manning. "There Must be a Better Way to Treat Gifted Researchers." *raum & zeit* (now *Explore!*) Vol. 2 No. 1 (1990): 2–3.

Kelly, Don. *The Manual of Free Energy Devices and Systems*. Clayton, GA: Cadake Industries and Copple House, 1987.

Kossy, Donna. *Kooks*. Portland, OR: Feral House, 1994.

Kovac, Ronald J. "Motion of Plasma as a Source of New Energy and Matter Transformation Emperical

Results." In *Proceedings of the International Symposium on New Energy*, in Denver, 12–15 May 1994. Fort Collins, CO: Rocky Mountain Research Institute, 1994, 271–282.

Krieger, Erwin. Letter to author, 21 June 1994.

Lindemann, Peter A. *A History of Free Energy Discoveries*. Bayside, CA: Borderland Sciences Research Foundation, 1986.

Mann, W. Edward, and Edward Hoffman. *The Man Who Dreamed of Tomorrow*. Los Angeles: J.P. Tarcher, 1980.

Manning, Jeane. "The Burial of Living Technology." In *Suppressed Inventions and Other Discoveries*, ed. Jonathan Eisen, 251–265. Auckland, New Zealand: Auckland Institute of Technology Press, 1994.

Manning, Jeane. "Gunfire in the Laboratory." In *Suppressed Inventions and Other Discoveries*, ed. Jonathan Eisen, 226–240. Auckland, New Zealand: Auckland Institute of Technology Press, 1994.

Moore, Clara Bloomfield. *Keely and His Discoveries*. Secaucus, NJ: University Books, 1893.

Moray, John E. "Radiant Energy." In *Proceedings of the First International Symposium on Non-Conventional Energy*, in Toronto, 23–24 October 1981. Toronto: George Hathaway, 1981, 316–319.

Moray, T. Henry. *Radiant Energy*. Bayside, CA: Borderland Sciences Research Foundation, 1945.

Moray, T. Henry, and John E. Moray. *The Sea of Energy*. Salt Lake City, UT: Cosray Research Institute, 1978.

Neill, A.S. "The Man Reich." In *Wilhelm Reich: The Evolution of His Work*, by David Boadella, appendix. London: Arkana, 1985.

Pond, Dale, ed. *Universal Laws Never Before Revealed: Keely's Secrets*. Santa Fe, NM: The Message Company, 1995.

Simmonds, C. Warren. "On the Subject of Radiant Energy." in *The Sea of Energy*, by T. Henry Moray and John E. Moray, 264–265. Salt Lake City, UT: Cosray Research Institute, 1978.

Simplified Technology Service. *Space Energy Receivers*. Bradley, IL: Lindsay Publications, 1984.

Skilling, Ed. Interview by author. Milwaukee, WI. Tape recording. 17 July 1993.

Chapter 4   A New Physics for a New Energy Source

Aspden, Harold. *Physics Without Einstein*. Southampton, England: Sabberton Publications, 1969.

Aspden, Harold. *Physics Unified*. Southampton, England: Sabberton Publications, 1980.

Bearden, Thomas. *Gravitobiology*. Chula Vista, CA: Tesla Book Company, 1991.

Bearden, Thomas, and Andrew Michrowski. *The Emerging Energy Science: An Annotated Bibliography*. Ottawa, Ont.: Planetary Association for Clean Energy, 1985.

Cole, Daniel, and Harold Puthoff. "Extracting Energy and Heat From the Vacuum." *Physical Review E* Vol. 48 No. 2 (August 1993): 1562–1565.

Davies, Owen. "Volatile Vacuums." *Omni*, February 1991, 50–56.

Essen, L. "Relativity—Joke or Swindle?" *Electronics and Wireless World* Vol. 94 No. 1624 (February 1988): 126–127.

Fox, Hal. "Space Energy—Peer Reviewed." *New Energy News*, February 1994, 2–6.

Fox, Hal. "Cold Nuclear Fusion, Space Energy Devices and Commercialization." In *Proceedings of the International Symposium on New Energy*, in Denver, 12–15 May 1994. Fort Collins, CO: Rocky Mountain Research Institute, 1994, 121–136.

Graneau, Peter. "Concept of a Capillary Fusion Reactor." In *Proceedings of the International Symposium on New Energy*, in Denver, 16–18 April 1993. Fort Collins, CO: International Association for New Science, 1993, 153–168.

Graneau, Peter. "Capillary Fusion." *Cold Fusion*, July/August 1994, 57–60.

Graneau, Peter, and Neal Graneau. *Newton Versus Einstein: How Matter Interacts With Matter*. New York: Carlton Press, 1993.

Haisch, Bernard, Alfonso Rueda, and Harold Puthoff. "Inertia as a Zero-Point Field Lorentz Force." *Physical Review A* Vol. 49 No. 2 (February 1994): 678–694.

Inomata, Shiuji, and Yoshiyuki Mita. "Design Considerations for Super-Conducting N-Machine." In *Proceedings of the International Symposium on New Energy*, in Denver, 12–15 May 1994. Fort Collins, CO: Rocky Mountain Research Institute, 1994, 199–218.

King, Moray B. *Tapping the Zero-Point Energy*. Provo, UT: Paraclete Publishing, 1989.

Kostro, L. "Einstein's New Conception of the Ether." *raum & zeit* (now *Explore!*) Vol. 2 No. 5 (1991): 81–84.

Motz, Lloyd, and Jefferson Hane Weaver. *The Story of Physics*. New York: Avon Books, 1989.

The New Illustrated Science and Invention Encyclopedia. Westport, CT: H.S. Stuttman Inc., 1989.

Nieper, Hans A. *Dr. Nieper's Revolution in Technology, Medicine and Society*. Oldenburg, Germany: MIT Verlag, 1985. Edgar Mitchell is quoted on p. 61.

Puthoff, Harold. "Ground State of Hydrogen as a Zero-point-fluctuation-determined State." *Physical Review D* Vol. 35 No. 10 (May 1987): 3266–3269.

Puthoff, Harold. "Everything for Nothing." *New Scientist* Vol. 127 No. 1727 (July 1990): 52–55.

Puthoff, Harold. "Zero Point Energy." *Fusion Facts* Vol. 3 No. 3 (September 1991): 1–2.

Puthoff, Harold. "Quantum Fluctuations of Empty Space: A New Rosetta Stone of Physics?" *Frontier Perspectives* Vol. 2 No. 2 (Fall/Winter 1991): 19–23.

Silvertooth, E.V. "Special Relativity." *Nature* Vol. 332 No. 6080 (1986): 590.

Silvertooth, E.V. "Experimental Detection of the Ether." *Speculations in Science and Technology* Vol. 10 No. 3 (1987): 3–7.

Tewari, Paramahamsa. *Beyond Matter*. Lekh Raj Nagar, India: Printwell Publications, 1984.

Chapter 5 Solid-State Energy Devices and Their Inventors

Davies, Owen. "Volatile Vacuums." *Omni*, February 1991, 50–56.

Hathaway, George. "The Hutchison Effect." *Electric Spacecraft Journal* Vol. 1 No. 4 (1991): 6–12.

Lambertson, Wingate. "A Constructive Role for Environmentally Concerned Citizens." *raum & zeit* (now *Explore!*) Vol. 1 No. 6 (1990): 84.

Lambertson, Wingate. "Electric Power from the Vacuum." *Explore!* Vol. 3 No. 5 (1992): 64–68.

Lambertson, Wingate. "History and

Status of the WIN Process." In *Proceedings of the International Symposium on New Energy*, in Denver, 12–15 May 1994. Fort Collins, CO: Rocky Mountain Research Institute, 1994, 283–288.

Manning, Jeane. "Rainbow in the Lab: The John Hutchison Story." *Electric Spacecraft Journal* Vol. 1 No. 4 (1991): 13–22.

Puthoff, Harold. "Quantum Fluctuations of Empty Space: A New Rosetta Stone of Physics?" *Frontier Perspectives* Vol. 2 No. 2 (Fall/Winter 1991): 19–23.

Shoulders, Kenneth R. U.S. Patent No. 5,018,180. 21 May 1991. *Energy Conversion Using High Charge Density.*

Chapter 6 Floyd Sweet— Solid-State Magnet Pioneer

Manning, Jeane. "New Energy Tech Announced at Prestigious Conference." *raum & zeit* (now *Explore!*) Vol. 3 No. 1 (1991): 69–75.

Margolin, Alvin R. "A Eulogy for Floyd `Sparky' Sweet." Desert Hot Springs, CA, 30 August 1995.

Rosenthal, Walt. "Floyd Sweet's VTA Unit." *Space Energy Newsletter* Vol. 4 No. 1 (March 1993): 1–4.

Silvertooth, E.V. "Special Relativity." *Nature* Vol. 332 No. 6080 (1986): 590.

Sweet, Floyd A. *Nothing is Something: The Theory and Operation of a Phase-Conjugated Vacuum Triode.* 24 June 1988.

Sweet, Floyd A. "The Vacuum Triode Amplifier." In *Free Energy: Final Solutions*, ed. Don Kelly. Clearwater, FL, 1990. Photocopy.

Sweet, Floyd, and Thomas Bearden. "Utilizing Scalar Electromagnetics to Tap Vacuum Energy." In *Proceedings*

of the 26th Intersociety Energy Conversion Energy Conference, in Boston, 4–9 August 1991. LaGrange, IL: American Nuclear Society, 1991, Vol. 4 No. 1: 370–375.

Watson, Michael. "Construction of Floyd Sweet's VTA." In *Proceedings of the International Symposium on New Energy*, in Denver, 12–15 May 1994. Fort Collins, CO: Rocky Mountain Research Institute, 1994, 435–444.

Watson, Michael. "The Status of the VTA." *New Energy News*, November 1994, 6.

Chapter 7 Rotating-Magnet Energy Innovators

DePalma, Bruce E. "Studies on Rotation Leading to the N-Machine." In *Proceedings of the First International Symposium on Non-Conventional Energy Technologies*, in Toronto, 23–24 October 1981. Toronto: George Hathaway, 1981, 247–258.

DePalma, Bruce E. "On the Possibility of Extraction of Electrical Energy Directly From Space." *Magnets in Your Future* Vol. 5 No. 8 (August 1991): 25–26.

DePalma, Bruce E. Editorial. *Space Energy Newsletter* Vol. 3 No. 3 (October/November 1992): 1–2.

Fox, Hal, Toby Grotz, and Andrew Michrowski. "The Denver Report." *Planetary Association for Clean Energy* Vol. 7 No. 4 (1994): 9–12.

How to Generate Electricity Without Consuming Any Fuel: Collected Papers of Dr. Bruce DePalma and Dr. Paramahamsa Tewari. Cedar Key, FL: People's Network Inc., 1990.

Inomata, Shiuji, and Yoshiyuki Mita. "Design Considerations for Super-Conducting N-Machine." In *Proceedings of the International Symposium on New Energy*, in Denver, 12–15 May 1994. Fort Collins, CO: Rocky

Mountain Research Institute, 1994, 199–218.

Kelly, Don. *The Manual of Free Energy Devices and Systems.* Clayton, GA: Cadake Industries and Copple House, 1987.

Knoll, Ernst. "A Motor Using Only Permanent Magnets?" *Untapped Technology in Review* Vol. 1 No. 1 (1994): 39.

Schaffranke, Rolf. "The Development of Post-Relativistic Concepts in Physics and Advanced Technology Abroad." In *Proceedings of the First International Symposium on Non-Conventional Energy Technologies,* in Toronto, 23–24 October 1981. Toronto: George Hathaway, 1981. Werner Heisenberg is quoted on page 287.

Tewari, Paramahamsa. *Beyond Matter.* Lekh Raj Nagar, India: Printwell Publications, 1984.

Tewari, Paramahamsa. "Generation of Cosmic Energy and Matter From Absolute Space." In *Proceedings of the International Symposium on New Energy,* in Denver, 16–18 April 1993. Fort Collins, CO: International Association for New Science, 1993, 219–303.

Valone, Thomas. "The Real Story of the N-Machine." *Extraordinary Science,* April/June 1994, 5–13.

Werjefelt, Bertil. "Energy From Magnetic Materials/Magnetic Fields." Paper presented at Cambridge, MA, 21 January 1995.

Werjefelt, Bertil. "The Magnetic Battery." *Electrifying Times,* Winter 1995, 2.

Chapter 8   *Cold Fusion—A Better Nuclear Technology*

Bockris, John O'Malley. Interview by author. Stanford University, CA. Tape recording. 11 August 1990.

Bockris, John O'Malley. Talk given at Institute for New Energy, Denver, April 1993.

Browne, Malcolm W. "New Shot at Cold Fusion by Pumping Sound Waves Into Tiny Bubbles." *The New York Times,* 20 December 1994.

Deak, David. "We Now Have New Physics." *Cold Fusion,* July/August 1994, 70–73.

Fox, Hal. *Cold Fusion Impact in the Enhanced Energy Age.* Salt Lake City, UT: Fusion Information Center, 1992.

Fox, Hal. "Scientists of the Year." *Fusion Facts* Vol. 3 No. 7 (January 1992): 2.

Fox, Hal. "Cold Nuclear Fusion, Space Energy Devices and Commercialization." In *Proceedings of the International Symposium on New Energy,* in Denver, 12–15 May 1994. Fort Collins, CO: Rocky Mountain Research Institute, 1994, 121–136.

Fox, Hal. "First Patent Issued." *New Energy News,* September 1994, 3–4.

Green, Wayne. "But Is It Real?" *Cold Fusion,* May 1994, 8.

Hodgkinson, Neville. "Storm in a Bucket." *London Sunday Times,* 27 June 1993.

Mallove, Eugene F. *Fire From Ice: Searching for the Truth Behind the Cold Fusion Furor.* New York: John Wiley & Sons, Inc., 1991.

Mallove, Eugene F. "Cold Fusion Goes Commercial." *Infinite Energy* Vol. 1 No. 2 (1995): 3–4.

Mallove, Eugene F. "Ignition! We Have Lift-Off!" *Infinite Energy* Vol. 1 No. 4 (1995): 3–4.

The New Illustrated Science and Invention Encyclopedia. Westport, CT: H.S. Stuttman Inc. Publishers, 1989.

Pollack, Andrew J. "Cold Fusion, Derided in U.S., Is Hot in Japan." *The New York Times,* 17 November 1992.

Rothwell, Jed, and Eugene Mallove. "A Cold Fusion Primer." *Cold Fusion*, May 1994, 49–54.

Silber, Kenneth. "Fusion Forces Hot Reactions." *Insight*, 14 March 1994, 14–16.

Storms, Edmund. "A Very Unscientific and Personal History of the Cold Fusion Effect." *New Energy News*, January 1994, 12.

Storms, Edmund. "Chemically-Assisted Nuclear Reactions." *Cold Fusion*, July/August 1994, 42–53.

Tinsley, Chris. "Only the Cold Fusion Critics Were Icarus." Review of *Too Close to the Sun*, BBC-TV Horizon in collaboration with CBC-TV. *Cold Fusion*, July/August 1994, 21.

Chapter 9   *Powering Up on Hydrogen*

Billings, Roger E. *The Hydrogen World View*. Independence, MO: International Academy of Science, 1991.

Bockris, John O'Malley. Interview by author. Stanford University, CA. Tape recording. 11 August 1990.

Bockris, John O'Malley, T. Nejat Veziroglu, and Debbi Smith. *Solar Hydrogen Energy: The Power to Save the Earth*. London: Macdonald & Co., 1991.

Day, James. *The Hindenburg Tragedy*. New York: The Bookwright Press, 1989.

Goldes, Mark. "On Demand Hydrogen Generator." *New Energy News*, May 1993, 8.

McNeil, Russell. "Search for the Fuel of the Future." *Canadian Geographic*, December 1989/January 1990, 114—118.

Tanaka, Shelley. *The Disaster of the Hindenburg: The Last Flight of the Greatest Airship Ever Built*. New York: Scholastic, Inc., 1993.

Westdyk, Karin. "U.S. Patent Granted to West Milford Resident for Unique Hydrogen Energy System." *The Messenger: Environmental Health Journal*, March/April 1992, 13–15.

Westdyk, Karin. "The Story of Francisco Pacheco and the Suppression of Hydrogen Technology." In *Suppressed Inventions and Other Discoveries*, ed. Jonathan Eisen, 343–346. Auckland, New Zealand: Auckland Institute of Technology Press, 1994.

Chapter 10   *Turning Waste Heat Into Electricity*

Aspden, Harold. "Magnets and Gravity." *Magnets in Your Future*, Vol. 6 No. 6 (June 1992): 15–22.

Aspden, Harold. "Electricity Without Magnetism?" *Electronics World and Wireless World* Vol. 98 No. 1674 (July 1992): 540–542.

Aspden, Harold. *Power From Magnetism*. Energy Science Report No. 1. Southampton, England: Sabberton Publications, 1994.

Aspden, Harold. "Magneto-Thermodynamics: A Progress Report." *New Energy News*, September 1994, 1–3.

Aspden, Harold. "Extracting Energy From a Magnet." *New Energy News* Vol. 3 No. 3 (August 1995): 1–3.

Aspden, Harold. "Power From Room Heat." *Nexus* Vol. 2 No. 27 (August/September 1995): 54.

The Concise Columbia Encyclopedia. 2nd ed. New York: Columbia University Press, 1989, 823.

Lindemann, Peter A. "Thermodynamics and Free Energy." *Borderlands* Vol. L No. 3 (Fall 1994): 6–10.

Tesla, Nikola. "The Problem of Increasing Human Energy." *The Century Illustrated Monthly Magazine*, June 1900, 210.

Wiseman, George. *The Negawatt.* Yahk, B.C.: Eagle Research Inc., 1992.

Wiseman, George. *Energy Conserver Book 1.* Yahk, B.C.: Eagle Research Inc., 1993.

Chapter 11   *Low-Impact Water Power—A New Twist on an Old Technology*

Alexandersson, Olof. *Living Water: Viktor Schauberger and the Secrets of Natural Energy.* Wellingborough, England: Turnstone Press Ltd., 1982.

British Columbia Hydro Power and Authority. *Davis Turbine Queen Charlotte Islands Tidal Power Study, Juskatla Narrows Overview Report.* Vancouver, B.C.: B.C. Hydro, July 1984.

British Columbia Hydro Power and Authority. *Environmental and Socio-Economic Overview. Queen Charlotte Islands Davis Turbine Tidal Power Project.* Vancouver, B.C.: B.C. Hydro, August 1985.

Coats, Callum. "The Magic and Majesty of Water: The Natural Eco-technological Theories of Viktor Schauberger." *Nexus* Vol. 2 No. 14 (June/July 1993): 36–39.

Curry, Stacey, and Shaligram Pokharel. "Micro Hydroelectricity in Nepal." *Alternatives* Vol. 19 No. 2 (1993): 6.

Deudney, Daniel, and Christopher Flavin. *Renewable Energy: The Power to Choose.* New York: W.W. Norton & Company, 1983.

Halvorson, Harold N. *Evaluation of Nova Energy Ltd's Hydro Turbine.* Victoria, B.C.: Halvorson Consultants Ltd, 9 December 1995. Unpublished paper written for the British Columbia Ministry of Employment and Investment.

Hume, Stephen. "Mega-project Mania Threatens to Sink Turbine Power." *Vancouver Sun,* 5 August 1994.

Ignazio, Joseph L. Unpublished open letter by Director of Planning, U.S. Army Corps of Engineers, New England Division, describing the Davis Hydro Turbine. 22 February 1994.

Nova Energy Ltd. press release. Vancouver, B.C., 11 September 1994.

Pratte, Bruce D. *General Comments on Nova Energy Ltd. Davis Hydro Turbine.* 23 February 1994. Unpublished paper written for National Research Council, Ottawa.

"Tesla Engine Builders Association." *Planetary Association for Clean Energy* Vol. 7 No. 4 (1994): 4.

Chapter 12   *The World of Power Possibilities*

China Railroad Corporation. *Testing Report of Eco-Kat.* 18 December 1993. Translated by Sally Lee. Unpublished report.

Davidson, Dan A. *Energy: Breakthroughs to New Free Energy Devices.* Sierra Vista, AZ: RIVAS, 1990.

Grander, Johann. *Wasserbelebung.* Seefeld, Austria: Umwelt Vertriegs Organization, 1994.

Grotz, Toby. "Institute for New Energy Trip Report: Around the World in 30 Days." *New Energy News,* January 1994, 5–11.

Huber, Georg. Letter to author, 6 November 1995.

Kelly, Don. Letters to author, 1988 through 1994.

Kronberger, Hans. "A Search for Clues." Translated by Zuzana Krkoskova. *Sonnen Zeitung* [*Sun Newspaper*] (Vienna), June 1994, 28–30.

Kronberger, Hans, and Siegbert Lattacher. *On the Track of Water's Secret.* Vienna: Uranus Verlagsgesellschaft, 1995.

Manning, Jeane. "Society for Scientific Exploration Airs Research on Anomalies." *raum & zeit* (now *Explore!*) Vol. 2 No. 5 (1991): 79–80.

Methernitha, letter to Don Kelly, 1988.

Nieper, Hans A. *Dr. Nieper's Revolution in Technology, Medicine and Society.* Oldenburg, Germany: MIT Verlag, 1985.

Proceedings of the Swiss Association for Free Energy Symposium, in Einsiedeln, Switzerland, 27–29 October 1989. Einsiedeln, Switzerland: Swiss Association for Free Energy, 1989.

Schaffranke, Rolf. *Ether-Technology.* Clayton, GA: Cadake Industries and Copple House, 1977.

Thesta-Distatica. Produced and directed by Methernitha. 30 minutes. Linden, Switzerland, 1989. Videocassette.

Chapter 13 *Harassing the Energy Innovators*

Andrews, Edmund L. "Cold War Secrecy Still Shrouds Inventions." *The New York Times,* 23 May 1992.

Brown, Paul. Unpublished "Open Letter to All Working on Alternate Energy." Oregon, 1 November 1991.

Davidson, John. *The Secret of the Creative Vacuum.* Essex, England: C.W. Daniel Co. Ltd., 1989.

Eisen, Jonathan, ed. *Suppressed Inventions and Other Discoveries.* Auckland, New Zealand: Auckland Institute of Technology Press, 1994.

"Electrogravitics Developments." *Planetary Association for Clean Energy* Vol. 7 No. 4 (1994): 7–8.

Gyorki, John R. "Losing a Battle Against Not-Invented-Here." *Machine Design,* 23 February 1989, 4.

Hasslberger, Josef. "A New Awareness." *raum & zeit* (now *Explore!*) Vol. 3 No. 1 (1991): 65–67.

King, Moray B. *Tapping the Zero-Point Energy.* Provo, UT: Paraclete Publishing, 1989.

LaViolette, Paul A. Remarks made at the Intersociety Energy Conversion Engineering Conference in Boston, 4–9 August 1991.

LaViolette, Paul A. "The U.S. Antigravity Squadron." In *Proceedings of the International Symposium on New Energy,* in Denver, 16–18 April 1993. Fort Collins, CO: International Association for New Science, 1993, 469–486.

MacNeill, Ken. "Insights Into the Proprietary Syndrome." In *Proceedings of the Second International Symposium on Non-Conventional Energy Technology,* in Atlanta, 9–11 September 1984. Winter Haven, FL: Cadake Industries, 1984, 125–126.

Manning, Jeane. "Anti-Gravity on the Rocks: The T.T. Brown Story." In *Suppressed Inventions and Other Discoveries,* ed. Jonathan Eisen, 267–277. Auckland, New Zealand: Auckland Institute of Technology Press, 1994.

Marinov, Stefan. Letter to Richard von Weizäcker. 10 October 1992.

Trombly, Adam. "Philosophical Overview of Free Energy." In *Proceedings of the Second International Symposium on Nonconventional Energy Technology,* in Atlanta, 9–11 September 1984. Winter Haven, FL: Cadake Industries, 1984, 143–144.

Chapter 14 *Society and a Free-Energy Economy*

Becker, Robert O. *Cross Currents.* Los Angeles: J.P. Tarcher, Inc., 1990.

Bell, Clare. "Petro$$$ Fund Carnegie Mellon Study." *Electrifying Times*, Fall 1995, 10, 18.

Bertell, Rosalie. "Exposing the Agenda of the Military Establishment." *Ecodecision*, September 1993, 82.

Bockris, John O'Malley. Interview by author. Stanford University, CA. Tape recording. 11 August 1990.

Flavin, Christopher, and Nicholas Lenssen. *Power Surge: Guide to the Coming Energy Revolution*. New York: W.W. Norton & Co., 1994.

Fox, Hal. *Cold Fusion Impact in the Enhanced Energy Age*. Salt Lake City, UT: Fusion Information Center, 1992.

Froning, H.D. Jr. "An Interstellar Exploration Initiative For Future Flight." *Space Technology* Vol. 13 No. 5 (1993): 503–512.

Jacobs, Michael. *The Green Economy*. Vancouver, B.C.: University of British Columbia Press, 1993.

Kovac, Ron J. "Plasma Shaping Reveals New Atomic Transformation Technique and Cold Fusion at Chemical-Molecular Levels." *Fulcrum—The Science Journal of the University of Science and Philosophy* Vol. 3 No. 2 (December 1994): 19.

Krieger, Erwin. Letter to author, 2 May 1994.

Lambertson, Wingate. *True Expenditure on Energy*. 25 September 1995.

Manning, Jeane. "New Energy Istitute: A Leap Into the Future." *Explore!* Vol. 4 No. 5 (1993): 79–83.

Mendillo, M., and others. "Spacelab-2 Plasma Depletion Experiments for Ionospheric and Radio Astronomical Studies." *Science* Vol. 238 No. 4831 (27 November 1987): 1260–1264.

Moore, Curtis, and Alan Miller. *Green Gold: Japan, Germany, the United States and the Race for Environmental Technology*. Boston: Beacon Press, 1994.

O'Leary, Brian. *Miracle in the Void*. Kihei, HI: Kamapua'a Press, 1996.

O'Rourke, P.J. *All the Trouble in the World: The Lighter Side of Overpopulation, Famine, Ecological Disaster, Ethnic Hatred, Plague and Poverty*. Toronto: Random House of Canada, 1994.

Peterson, John L. *The Road to 2012: Looking Toward the Next Two Decades*. Arlington, VA: Arlington Institute, 1992.

Proceedings of the Swiss Association for Free Energy Symposium, in Einsiedeln, Switzerland, 27–29 October 1989. Einsiedeln, Switzerland: Swiss Association for Free Energy, 1989.

Ronan, Colin A., ed. *Science Explained*. New York: Henry Holt & Co., 1993.

Strauss, Stephen. "NASA Cloud Program will Paint Up the Sky for Most Canadians." *Globe and Mail*, 22 March 1989.

U.S. Bureau of the Census. "State Government Tax Collections and Excise Taxes, by State: 1992." *Statistical Abstract of the United States*. Washington, DC: Bureau of the Census, 1994.

Chapter 15   The Power Is in Our Hands

Bearden, Thomas. "Overunity Electrical Power Efficiency Using Energy Shuttling Between Two Circuits." In *Proceedings of the International Symposium on New Energy*, in Denver, 12–15 May 1994. Fort Collins, CO: Rocky Mountain Research Institute, 1994, 46–66.

Bird, Christopher. "The Saga of Yull Brown." *Explore!* Vol. 3 (1992): No. 2, 49–66; No. 3, 58–66; No. 6, 47–62 and *Explore More!* Vol. 1 (1995): No. 10, 55–63; No. 12, 19–24, 56–60; No. 13, 43–49.

Callenbach, Ernest. *Ecotopia Emerging.* Berkeley, CA: Banyan Tree Books, 1981.

Davidson, John. *The Secret of the Creative Vacuum.* Essex, England: C.W. Daniel Co. Ltd., 1989.

Davies, Owen. "Volatile Vacuums." *Omni,* February 1991, 50–56.

Fox, Hal. *Cold Fusion Impact in the Enhanced Energy Age.* Salt Lake City, UT: Fusion Information Center, 1992.

Fox, Hal, Toby Grotz, and Andrew Michrowski. "The Denver Report." *Planetary Association for Clean Energy* Vol. 7 No. 4 (1994): 9–12.

Goldes, Mark. "Takahashi Motor Released." *New Energy News,* January 1996, 8.

Henderson, Hazel. *Paradigms in Progress: Life Beyond Economics.* Indianapolis, IN: Knowledge Systems, Inc., 1992.

Jensen, Paul R. "The Unidirectional Transformer." In *Proceedings of the International Symposium on New Energy,* in Denver, 12–15 May 1994. Fort Collins, CO: Rocky Mountain Research Institute, 1994, 545–550.

LaViolette, Paul. "The U.S. Antigravity Squadron." In *Proceedings of the International Symposium on New Energy,* in Denver, 16–18 April 1993. Fort Collins, CO: International Association for New Science, 1993, 469–486.

Manning, Jeane. "Magnet Motor Researcher's Quest." *Extraordinary Science,* January/March 1990, 8–14.

McClain, Joel, and Norman Wooten. "The Magnetic Resonance Amplifier." *Electrifying Times,* Winter/Spring 1995, 3.

Puthoff, Harold. "Quantum Fluctuations of Empty Space: A New Rosetta Stone of Physics?" *Frontier Perspectives* Vol. 2 No. 2 (Fall/Winter 1991): 19–23.

Rubik, Beverly. "Science: a Feminine Perspective." *Creation,* November/December 1990, 6–7.

Schuster, Michael L. *Continuous Energy.* Milwaukee, WI: Sufra Publications, 1990.

Sweet, Floyd, and Thomas Bearden. "Utilizing Scalar Electromagnetics to Tap Vacuum Energy." In *Proceedings of the 26th Intersociety Energy Conversion Energy Conference,* in Boston, 4–9 August 1991. LaGrange, IL: American Nuclear Society, 1991, Vol. 4 No. 1: 370–375.

"Trees Tune Into Radio Waves." *Nexus,* June/July 1995, 7.

Winter, Daniel, and others. *Alphabet of the Heart: Sacred Geometry.* Eden, NY: Crystal Hill Farm, 1992.

Index

740327